Radially Symmetric Patterns of Reaction-Diffusion Systems

MEMOIRS
of the
American Mathematical Society

Number 786

Radially Symmetric Patterns of Reaction-Diffusion Systems

Arnd Scheel

September 2003 • Volume 165 • Number 786 (third of 4 numbers) • ISSN 0065-9266

American Mathematical Society
Providence, Rhode Island

2000 *Mathematics Subject Classification.*
Primary 35K57, 35B32, 37L10, 37L15, 34C37; Secondary 35J60, 35B40, 37G40.

Library of Congress Cataloging-in-Publication Data

Scheel, Arnd, 1966–
Radially symmetric patterns of reaction-diffusion systems / Arnd Scheel.
p. cm. — (Memoirs of the American Mathematical Society, ISSN 0065-9266 ; no. 786)
"Volume 165, number 786 (third of 4 numbers)."
Includes bibliographical references.
ISBN 0-8218-3373-1 (alk. paper)
1. Reaction-diffusion equations. 2. Bifurcation theory. 3. Normal forms (Mathematics) I. Title. II. Series.

QA3.A57 no. 786
[QA377]
510 s—dc21
[515′.353] 2003051901

Memoirs of the American Mathematical Society

This journal is devoted entirely to research in pure and applied mathematics.

Subscription information. The 2003 subscription begins with volume 161 and consists of six mailings, each containing one or more numbers. Subscription prices for 2003 are $555 list, $444 institutional member. A late charge of 10% of the subscription price will be imposed on orders received from nonmembers after January 1 of the subscription year. Subscribers outside the United States and India must pay a postage surcharge of $31; subscribers in India must pay a postage surcharge of $43. Expedited delivery to destinations in North America $35; elsewhere $130. Each number may be ordered separately; *please specify number* when ordering an individual number. For prices and titles of recently released numbers, see the New Publications sections of the *Notices of the American Mathematical Society.*

Back number information. For back issues see the *AMS Catalog of Publications.*

Subscriptions and orders should be addressed to the American Mathematical Society, P. O. Box 845904, Boston, MA 02284-5904, USA. *All orders must be accompanied by payment.* Other correspondence should be addressed to 201 Charles Street, Providence, RI 02904-2294, USA.

Memoirs of the American Mathematical Society is published bimonthly (each volume consisting usually of more than one number) by the American Mathematical Society at 201 Charles Street, Providence, RI 02904-2294, USA. Periodicals postage paid at Providence, RI. Postmaster: Send address changes to Memoirs, American Mathematical Society, 201 Charles Street, Providence, RI 02904-2294, USA.

This publication is indexed in *Science Citation Index*®, *SciSearch*®, *Research Alert*®, *CompuMath Citation Index*®, *Current Contents*®/*Physical, Chemical & Earth Sciences.*
Printed in the United States of America.

∞ The paper used in this book is acid-free and falls within the guidelines established to ensure permanence and durability.
Visit the AMS home page at `http://www.ams.org/`

10 9 8 7 6 5 4 3 2 1 08 07 06 05 04 03

... für Yann, Marlen und Miriam

Contents

Abstract

In this paper, bifurcations of stationary and time-periodic solutions to reaction-diffusion systems are studied. We develop a center-manifold and normal form theory for radial dynamics which allows for a complete description of radially symmetric patterns. In particular, we show the existence of localized pulses near saddle-nodes, critical Gibbs kernels in the cusp, focus patterns in Turing instabilities, and active or passive target patterns in oscillatory instabilities.

Received by the editor February 15, 2001.

1991 *Mathematics Subject Classification.* Primary 35K57, 35B32, 37L10, 37L15, 34C37 ; Secondary 35J60, 35B40, 37G40.

Key words and phrases. reaction-diffusion systems, defects, target patterns, center manifolds, normal forms.

Arnd Scheel was supported in part by NSF Grant #DMS-0203301.

CHAPTER 1

Introduction

Pattern formation in reaction-diffusion systems has been an intensive area of research in pure and applied mathematics, in chemistry, in biology, and in physiology. A challenging variety of biological patterns is found in D'Arcy Thompson's work "On growth and form" from 1917, who was one of the first to attempt a mathematical, though still phenomenological description of pattern forming processes [**Tho17**]. Later, Alan Turing [**Tur52**] emphasized the crucial interplay between simply chemical reaction kinetics and diffusive transport in modeling pattern forming processes. Reducing the complex inner-cellular dynamics to a simple reaction-diffusion system, he was able to explain the creation of stable structured states from an unstructured, homogeneous equilibrium of the system. However, reaction-diffusion systems attracted mathematical interest much earlier. For example, in the early work by Kolmogorov, Petrovsky, and Piscounov [**KPP37**] on travelling waves, we find the first mathematical attempt to describe competition and interaction between different states of a chemical or biological system.

Spatio-temporal pattern formation became accessible to a systematic mathematical study through the development of bifurcation theory. Bifurcation theory, in a very broad sense, is the study of qualitative changes in the behavior of a dynamical system, caused by variations of certain control parameters. There are two typical examples. The simplest one is the fold, where in the equation

$$\dot{x} = \mu - x^2,$$

two new equilibria appear when the parameter μ is increased above zero. The mathematical analysis arguably goes back to ancient Babylonia 700 B.C., where the quadratic formula was first discovered. In the second, typical, instability mechanism in a dynamical system, a periodic solution is created when a stationary state looses its stability; see [**AW30, Hop43**]. In the complex model equation

$$\dot{z} = (\mu + \mathrm{i})z - z|z|^2 \in \mathbb{C},$$

a stable periodic orbit is created when the parameter μ is increased above zero and the origin looses stability; see the bifurcation diagrams in Figure 1. The fold, typically referred to as a saddle-node bifurcation in dynamical systems, and the Hopf bifurcation occur in open classes of one-parameter families of dynamical systems. In both cases, at most one dynamically stable state of the dynamical system exists for each fixed parameter value. Adding a second parameter, one can find bifurcations in typical families of dynamical systems, where two stable states coexist for fixed parameter values — a situation, we are particularly interested in, here. Typical examples are the cusp

$$\dot{x} = \mu_1 - \mu_2 x - x^3$$

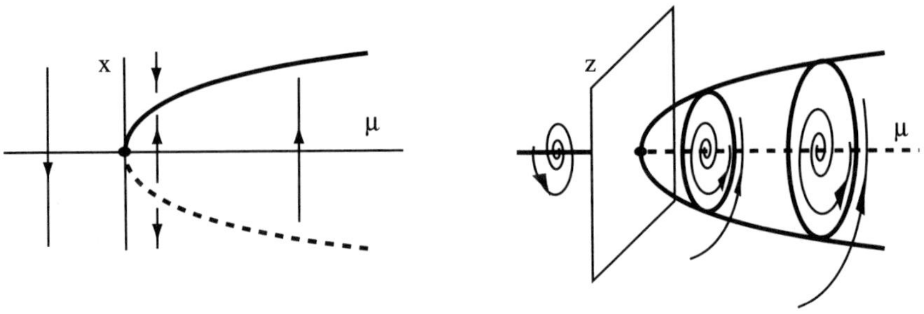

FIGURE 1. *Bifurcation diagrams for the fold and the Hopf bifurcation. Dashed lines denote unstable solutions, arrows indicate dynamics.*

and the weakly subcritical Hopf bifurcation

$$\dot{z} = (\mu_1 + \mathrm{i})z + \mu_2 z|z|^2 - z|z|^4 \in \mathbb{C},$$

see Figure 2 for the bifurcation diagrams.

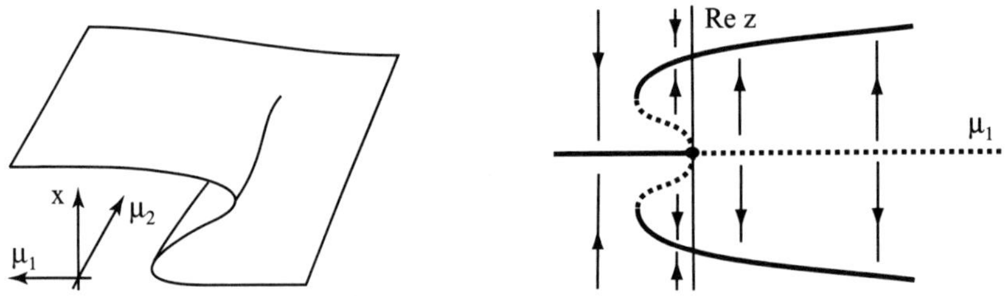

FIGURE 2. *The two-parameter bifurcation diagram for the cusp and the bifurcation diagram for the degenerate Hopf bifurcation in case* $\mu_2 > 0$.

When the state variable of a dynamical system depends on a spatial variable, many new questions arise. As already mentioned, Turing [**Tur52**] first noticed that spatially distributed systems exhibit different instability mechanisms. In a two-species reaction-diffusion system, the spatially homogeneous equilibrium may become unstable first with respect to spatially structured perturbations — despite the spatially homogenizing effect of diffusive coupling. Nonlinear saturation may then lead to the creation of steady, spatially periodic, stable patterns, which are usually referred to as Turing patterns (although his analysis was purely linear). The example in Turing's work was a (linear and spatially discrete) prototype of an activator-inhibitor system

$$\begin{aligned} \partial_t U_1 &= \partial_{xx} U_1 + f(U_1, U_2) \\ \partial_t U_2 &= d\partial_{xx} U_2 + g(U_1, U_2), \end{aligned}$$

where $d > 1$, $f(0,0) = g(0,0) = 0$, $\partial_{U_1} f > 0$, $\partial_{U_1} g > 0$ and $\partial_{U_2} f < 0$, $\partial_{U_2} g < 0$. More generally, instabilities in reaction-diffusion systems

$$\partial_t U = D\Delta_x U + F(U; \mu) \tag{1}$$

lead to the emergence of families of spatio-temporally periodic states, parameterized by the spatial wave number k close to a critical wave number $|k| \sim k_* \geq 0$. Bifurcation theory, through the steps of reduction to center manifolds, followed by normal form transformation and discussion of the reduced equations — possibly exploiting various symmetries of the problem — was particularly successful in explaining shape and dynamics of spatially periodic patterns such as stripes or hexagons [**GSS88, IA92, CI94**]. However, localized structures like defects or interfaces between spatially homogeneous or periodic states escaped these approaches which impose spatial periodicity in function spaces.

A second, important, question, particular to spatially extended systems, arises when a stable and an unstable, or two different stable equilibria coexist. If spatial coupling is weak or, if the domain is very large, the system might initially be in two different states in different regions of the domain. Spatial competition between the two states is best described by the motion of interfaces. The most detailed results on existence and stability of these interfaces are available in one spatial dimension. Interfaces between spatially homogeneous equilibrium states often propagate with constant speed and are called travelling waves. They can be found as spatially structured equilibria of the original reaction-diffusion system in an appropriately comoving frame $\xi = x - ct$, in which the equation reads

$$0 = U_t = D\partial_{\xi\xi}U + c\partial_\xi U + F(U;\mu);$$

see Figure 3.

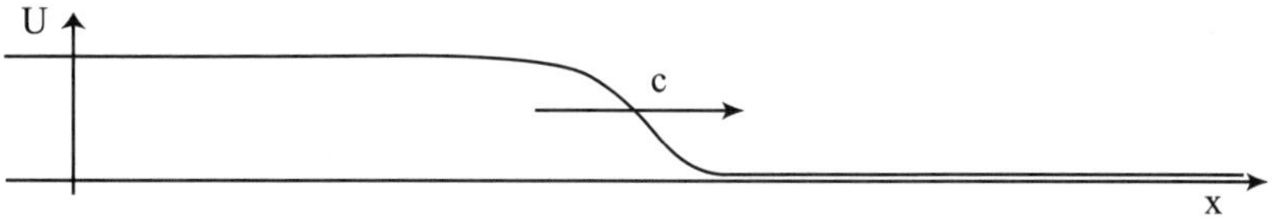

FIGURE 3. *A schematic picture of a travelling wave solution to a reaction diffusion system.*

This steady-state equation can then be rewritten as a dynamical system in spatial "time" ξ

$$U_\xi = V, \qquad V_\xi = -D^{-1}(cV + F(U;\mu)).$$

Travelling waves are heteroclinic orbits to the above ordinary differential equation, see Figure 4. However, interfaces between time-periodic patterns and homogeneous states cannot be found in the above steady-state ordinary differential equation. Similarly, competition between Turing patterns and spatially homogeneous equilibria leads to an oscillatory motion of the interface, and becomes invisible in the steady-state equation.

As a partial remedy, Kopell and Howard suggested to study oscillatory patterns in λ-ω-systems, where a rotational symmetry in the reaction kinetics enabled them to find oscillatory patterns as equilibria in a spatially comoving and kinetically corotating frame [**HK77**]. The extension from the artificial rotational equivariance to general kinetics remained formal in this early work; see [**DSSS00**] for a recent account.

Kirchgässner overcame these technical difficulties in a slightly different context [**Kir82**]. He could generalize center manifold theory to dynamically ill-posed

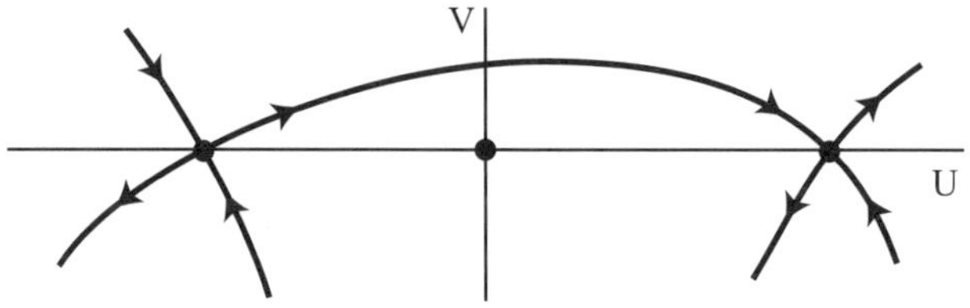

FIGURE 4. *The travelling wave in the phase space of the ODE.*

elliptic problems, as they arise in fluid mechanics or in the above mentioned oscillatory motion of interfaces. On the center manifold, he recovered an ordinary differential equation which described small amplitude solutions of an elliptic equation in an unbounded strip. The idea was largely exploited later in the study of free surface water-waves [**Kir88, IK92**], and in elasticity problems [**Mie88b**]. Morally, the emanation of small amplitude patterns is caused by essential spectrum of the linearized operator on the imaginary axis. Noncompact translation symmetry, unavoidable in an idealized description of competition by propagating interfaces, enforces failure of compactness of the resolvent and prevents the application of direct methods from bifurcation theory, such as Lyapunov-Schmidt reduction. The usage of spatial dynamics avoids the essential spectrum. In the restriction to steady or time-periodic patterns, the essential spectrum appears in the form of *isolated* neutral eigenvalues on the imaginary axis. For example, the continuum of eigenvalues $\lambda = -k^2$, $k \in \mathbb{R}$, of the operator ∂_{xx} on the real line turns into a simple Jordan block to the eigenvalue 0 for the steady-state problem $u_x = v$, $v_x = 0$. The nonlinear bifurcation problem for the spatial dynamical system is then accessible to methods from bifurcation theory — although dynamics now do have a completely different interpretation as spatial instead of temporal changes in the solution profile.

However, the approach via spatial dynamics is constrained to one effective unbounded dimension, singled out as the direction of spatial time. Attempts to generalize the approach to more than one space dimension lead to nonlocal bifurcation equations [**Mie92**].

A more general approach to bifurcations from the essential spectrum has been developed, exploiting variational structure or sign conditions on the nonlinearity instead of spatial dynamics; see the review in [**Stu97**]. In consequence, a detailed description of the bifurcation diagram could be obtained only in special cases.

The goal of the present work is to extend the method of spatial dynamics to higher space dimensions, restricting the attention to *radially symmetric* solutions. The radius serves as the spatial time variable.

The literature on radially symmetric solutions to elliptic problems is vast; see for example the references in [**Bre99**]. Nevertheless, the present work gives new contributions in at least three aspects.

First of all, we develop a systematic bifurcation theory, including the particular case of essential spectrum crossing the imaginary axis. Main result is a reduction method together with a normal form theorem. Nontrivial solutions are found from a matching procedure, replacing typical shooting arguments in previous works. The reduction procedure shows that, even for systems of equations, the typical phenomena from scalar elliptic equations in $\mathbb{R}^n$ occur, whenever the kernel of the linearization is one-dimensional, see Theorems 3.18 and 3.19 for the cases of a fold and a cusp in the reaction kinetics, respectively.

Second, in systems of elliptic equations, the Turing instability, which leads to stable, spatially periodic patterns in one space dimension, may create focus patterns; see Theorem 3.20. These patterns resemble stationary targets, where level lines of a fixed species come in infinitely many circles, spaced almost equidistantly. Along rays, emanating from the center, we find the one-dimensional, stable, spatially periodic structures. A variant of these patterns has been observed in Rayleigh-Bénard convection [**Cro89**], where a Turing-type instability leads to a convective state replacing the unstable pure conductive state of the fluid; see Figure 5. Level lines of a fixed species here correspond to regions of, say, upward pointing velocity fields. A (formal) theoretical analysis of such patterns has been initiated in [**BS78, PZM85**]. However, approximate descriptions using phase-diffusion equations break down in a neighborhood of the center of the pattern.

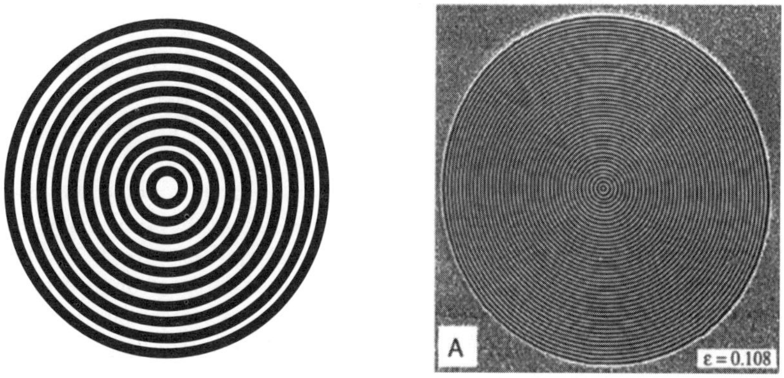

FIGURE 5. *A schematic plot of the bifurcating focus pattern and a picture from Rayleigh-Bénard convection in a cylindrical container; with permission, from the Annual Review of Fluid Mechanics, Volume 32, 2000 by Annual Reviews, www.AnnualReviews.org* [**BPA00**].

Thirdly, we are able to adapt the method to time-periodic patterns, including target patterns as observed in the Belousov-Zhabotinsky reaction; see Theorems 4.12 and 4.13 for existence of target patterns. We also refer to [**Gre78, KH81a, KH81b**] for previous work on existence of target patterns. Phenomenologically, a point-source in the center of the pattern is emitting waves in concentric circles. Far away from the center, we see spatially periodic waves along rays emanating from the center, which travel away from the center with positive phase and group velocity.

Technically, the main idea for the reduction method is inspired by the reduction procedure in [**Sche98**]. The goal there was to find spiral wave solutions in reaction-diffusion systems close to a Hopf bifurcation point in the reaction kinetics. In particular, solutions there were allowed to depend on the angular variable φ, whereas we allow dependence on time t, here. The technical difficulties, here, arise in the center of the patterns, whereas in [**Sche98**], the far-field required a more subtle analysis. Extending the results in [**Sche98**], we develop a normal form theory for dynamics in the radial variable, which allows us to treat general nonlinearities.

Physically, the patterns found in the present work should be interpreted partly as point defects, partly as coexistence boundaries. The pattern of concentric circles is a defect because the local wave vector cannot be extended continuously into the origin with range in the circular band of allowed unstable wave vectors. Coexistence patterns are found in case of a cusp in the kinetics. The kernel of the linearization in the kinetics is one-dimensional and we recover at leading order patterns from scalar elliptic equations. One of the stable states occupies a large disc or ball (the Gibbs kernel), whereas the rest of the domain is filled with the other stable pattern. Coexistence patterns occur in open regions of parameter space, but are typically unstable. They are most accurately interpreted as the minimally sized region, the inner pattern has to occupy in order to conquer the outer state. Smaller discs or balls will eventually disappear, whereas any larger disc will grow and tend to fill the whole domain. This is in contrast to the spatially one-dimensional situation, where coexistence between homogeneous or temporally periodic states only occurs for exceptional parameter values, as a codimension-one phenomenon. The reason is a type of interfacial energy which increases linearly (in $\mathbb{R}^2$) with the diameter of the disc and penalizes growth of the disc. The region filled with the preferred pattern, which is proportional to the energy gain by expansion, increases quadratically with the diameter. The patterns uncovered here mark precisely the point of balance between these two growth behaviors. We emphasize however that the dynamics we consider are not gradients of an energy functional, not even gradient-like. Gradient-like dynamics are restored only in the case of a stationary instability and only after reduction to a spatial center manifold. Coexistence patterns occur in oscillatory instabilities as well. In the weakly subcritical Hopf bifurcation, we may find a centered ball filled with the stable stationary state, whereas the complement is filled with concentric wave-trains, which are emitted from the homogeneous state. In this sense, we find coexistence as a new mechanism for the creation of target patterns.

Defects and critically sized interfaces as we find them here occur in many systems, other than reaction-diffusion systems. We mention phase-field models [**CF87**] from material science, Rayleigh-Bénard convection [**Cro89**], and the Swift-Hohenberg equation [**SH77**]. Oscillatory patterns are observed for example in chemical reactions [**FB85**], in nematic liquid crystals [**CH93**, Sec. IX.C], and in transversely extended lasers [**AGRR90, BBLPPTW91**]. Still, reaction-diffusion systems seem to provide a sufficiently large class of spatially extended dynamical systems, where *typical* spatio-temporal patterns can be analyzed systematically.

Outline: In Chapter 2, we review instabilities in one spatial dimension. The main technical tools such as center manifold reduction, normal form transformations, and transversality are introduced. We focus on coexistence patterns. Most of the results are not new, although they have not been stated for the case of reaction-diffusion systems in the literature. We mostly sketch the proofs, emphasizing the crucial points that will reappear in the analysis of radially symmetric patterns, later. In the subsequent chapters, we rely on the basic ideas introduced in this chapter. The main results are contained in Chapters 3 and 4.

In Chapter 3 we prove a nonautonomous center manifold reduction, Section 2, and give a normal form algorithm, Section 3. Scaling, derivation of universal reduced equations at leading order, and matching with the core region on the center manifold, as the main part of the bifurcation analysis, are found in Section 4.

Chapter 4 extends the reduction procedure to radially symmetric, time-periodic solutions. The reduction theorem is stated and proved in Section 2. Reduced vector fields and typical examples of patterns are given in Sections 3 and 4.

We conclude in Chapter 5 with an extensive discussion. In particular, we outline a stability analysis of radially symmetric patterns, we show existence of non radially symmetric patterns, and we briefly discuss the effects of a small hole in the center of the domain.

Acknowledgments: Many colleagues and friends have contributed to this work. First, I want to thank Bernold Fiedler for support and criticism throughout the years. Without his insistence, this paper would not have been written.

Large parts of the discussion on stability are motivated by joint work with Björn Sandstede during the past five years. Some of the ideas on coexistence derived benefit from a visit at the "Institut Non Linéaire de Nice-Sophia Antipolis". I very much enjoyed the exciting atmosphere in discussions with Pierre Coullet, Gérard Iooss, and Eric Lombardi.

I would also like to thank Arjen Doelman and Gérard Iooss for their many helpful comments on an earlier version.

To my colleagues in Berlin, Messoud Efendiev, Jörg Härterich, Christian Leis, Stefan Liebscher, and Karsten Matthies I owe many thanks for sharing their time and ideas with me.

Finally, I want to thank my family for their encouragement and support.

CHAPTER 2

Instabilities in one space dimension

1. Introduction

We consider reaction-diffusion systems

$$U_t = DU_{xx} + F(U;\mu), \tag{2}$$

with N species $U \in \mathbb{R}^N$, a p-dimensional real parameter $\mu \in \mathbb{R}^p$, smooth reaction kinetics $F \in C^\infty(\mathbb{R}^N \times \mathbb{R}^p, \mathbb{R}^N)$, and positive, diagonal diffusion matrix $D = \operatorname{diag}(d_j) > 0$, on the real line $x \in \mathbb{R}$. We are interested in pattern formation from a homogeneous equilibrium state. We therefore assume that $U(t,x) \equiv 0$ is a solution of (2) at $\mu = 0$. Aiming towards a bifurcation analysis, we consider the linearized equation about the equilibrium $U \equiv 0$:

$$V_t = DV_{xx} + \partial_U F(0;0)V =: \mathcal{L}_0 V. \tag{3}$$

Equations (2) and (3) define (local) semi-flows on the space of bounded, uniformly continuous functions $BC^0_{\text{unif}}(\mathbb{R}, \mathbb{R}^N)$. The operator $\mathcal{L}_0 : \mathcal{D}(\mathcal{L}_0) \subset BC^0_{\text{unif}} \to BC^0_{\text{unif}}$ is a sectorial operator [**Hen81, Yos71**], which we consider on the complexification $BC^0_{\text{unif}}(\mathbb{R}, \mathbb{C}^N)$ as well. Denote by $\operatorname{spec}\mathcal{L}_0$ the spectrum of $\mathcal{L}_0$. It is not hard to see that $U \equiv 0$ is isolated in the set of bounded solutions $\mathcal{I} \subset BC^0_{\text{unif}}(\mathbb{R}, \mathbb{R}^N)$ defined as

$$\mathcal{I} = \{U_0;\ \text{there is a solution } U(t,x) \in BC^0(\mathbb{R}\times\mathbb{R}, \mathbb{R}^N) \text{ of (2), with } U(0,\cdot) = U_0\},$$

if $\operatorname{spec}\mathcal{L}_0 \subset \mathbb{C}_- := \{z \in \mathbb{C}; \operatorname{Re} z < 0\}$; see [**Hen81**]. We therefore analyze the situation, when $\operatorname{spec}\mathcal{L}_0 \subseteq \overline{\mathbb{C}}_-$ but $\operatorname{spec}\mathcal{L}_0 \not\subset \mathbb{C}_-$.

In Section 2, we classify typical scenarios of marginal stability, where $\operatorname{spec}\mathcal{L}_0$ touches the imaginary axis. Upon varying the parameter μ, the spectrum of the linearization about the trivial equilibrium typically crosses the imaginary axis.

We then proceed to analyze bifurcating solutions in Sections 3 and 4, depending on whether $\operatorname{spec}\mathcal{L}_0 \cap \mathrm{i}\mathbb{R} = \{0\}$, a stationary instability, or $\operatorname{spec}\mathcal{L}_0 \cap \mathrm{i}\mathbb{R} = \{\pm\mathrm{i}\omega_*\}$, an oscillatory instability. The strategy will be to look for time-independent or time-periodic solutions.

We use center manifold reduction and normal form theory to reduce the problem to a universal model equation. Objects of interest are steady interfaces between stable states of the reaction-diffusion system.

Most of the results are not new, though one might not find the statements for the case of reaction-diffusion systems in the literature. Center manifold reduction goes back to [**Pli64, Kel67**], for ordinary differential equations, and to [**Kir82, Mie86, IM91**] for elliptic-type partial differential equations, as considered in Section 4.

Normal form theory as used here has been elaborated in [**ETBCI87**]; see also the review [**CS90**]. The discussion of solutions of model equations in the stationary case is elementary, whereas in the case of Hopf bifurcation, we refer to results of [**KH81a**] and [**vSH92**].

2. Classifying instabilities of reaction-diffusion systems

We start analyzing the spectrum of $\mathcal{L}_0$ as defined in (3). The following elementary lemma is well-known.

LEMMA 2.1. *A complex number λ belongs to the spectrum* $\operatorname{spec}\mathcal{L}_0$ *if, and only if, there is a $k \in \mathbb{R}$ such that*

$$d(k,\lambda) = \det\left(-Dk^2 + \partial_U F(0;0) - \lambda\right) = 0. \tag{4}$$

Proof. First assume the determinant is zero and let $U_0 \in \mathbb{C}^N$ belong to the kernel of $(-Dk^2 + \partial_U F(0;0) - \lambda)$. Then $U_0 e^{ikx}$ belongs to the kernel of $\mathcal{L}_0 - \lambda$, which proves the 'if'-part.

For the 'only-if'-part, let us first consider the operator $\mathcal{L}_0$ on $L^2(\mathbb{R}, \mathbb{C}^N)$. Then $\mathcal{L}_0$ is conjugate to its Fourier transform $\hat{\mathcal{L}}_0$ with $((\hat{\mathcal{L}}_0 - \lambda)\hat{U})(k) = -k^2 D\hat{U}(k) + \partial_U F(0;0)\hat{U}(k) - \lambda\hat{U}(k)$. Since the right side is invertible for each k, we obtain the inverse of $\mathcal{L}_0 - \lambda$ as the convolution with the Fourier transform of $(-Dk^2 + \partial_U F(0;0) - \lambda)^{-1}$, which is an exponentially localized function. The convolution operator can also be defined on $BC^0_{\mathrm{unif}}(\mathbb{R}, \mathbb{C}^N)$, where it defines a continuous operator $\tilde{\mathcal{L}}$, which can easily be checked to be a left inverse of $\mathcal{L}_0 - \lambda$. Since $\mathcal{L}_0 - \lambda$ is injective — elements in the kernel are solutions to an ordinary differential equation which does not have spectrum on the imaginary axis —, $\tilde{\mathcal{L}} - \lambda$ is also a right inverse, which proves the lemma. ∎

Summarizing, we may assume

$$\det\left(-Dk_*^2 + \partial_U F(0;0) - i\omega_*\right) = 0, \tag{5}$$

for some $k_* \in \mathbb{R}$, whenever we want to assume $i\omega_*$ belongs to $\operatorname{spec}\mathcal{L}_0$.

HYPOTHESIS 2.2. *[Criticality]* Assume $\mathcal{L}_0$ is marginally stable with critical eigenfunction $U_0 e^{ik_* x}$ to the unique critical eigenvalue $i\omega_*$, up to complex conjugation:

(i) $\det(-Dk^2 + \partial_U F(0;0) - \lambda) \neq 0$ for all $k \in \mathbb{R}$ and all $\lambda \in \mathbb{C}$ with $\lambda \neq i\omega_*$ and $\operatorname{Re}\lambda \geq 0$.
(ii) $\det(-Dk^2 + \partial_U F(0;0) - i\omega_*) \neq 0$ for all $k \neq \pm k_*$.
(iii) $\det(-Dk_*^2 + \partial_U F(0;0) - i\omega_*) = 0$.

DEFINITION 2.3. *[Types of instability]* According to Hypothesis 2.2, we distinguish four different cases of criticality:

(i) stationary homogeneous instabilities (O); $k_* = 0,\ \omega_* = 0.$
(ii) Turing instabilities (T); $k_* \neq 0,\ \omega_* = 0.$
(iii) Hopf instabilities (H); $k_* = 0,\ \omega_* \neq 0.$
(iv) Turing-Hopf instabilities (TH); $k_* \neq 0,\ \omega_* \neq 0.$

The first two cases are referred to as stationary instabilities, the other two as oscillatory instabilities.

The spatio-temporal shape of the eigenfunctions is illustrated in Figure 1.

REMARK 2.4. *Sometimes, the name Turing-Hopf bifurcation is used to refer to a codimension-two instability, where simultaneously two eigenfunctions, a Hopf*

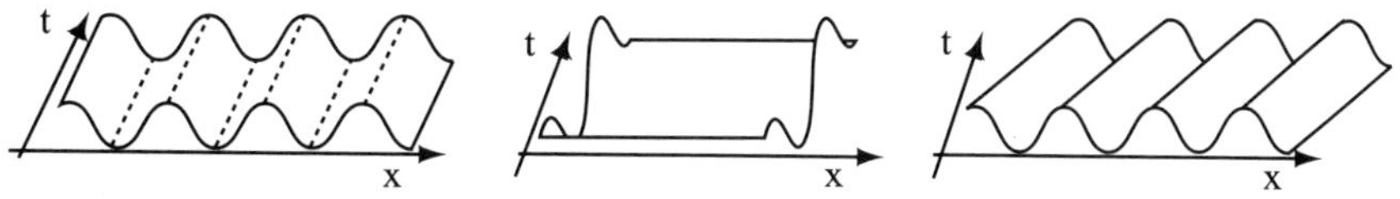

FIGURE 1. *Spatio-temporal shape of eigenfunctions in cases (T), (H), and (TH) from left to right.*

eigenfunction $U_1 e^{i\omega t}$ *and a Turing eigenfunction* $U_2 e^{ikx}$ *are critical. We do not discuss this type of instability, here.*

We assumed one, and only one, critical eigenvalue $i\omega_*$, up to complex conjugation, to be located in the closed right half plane. We also assume that this eigenvalue is simple in a general sense.

HYPOTHESIS 2.5. *[Simple critical eigenfunction]* We assume that

$$\partial_\lambda[\det\left(-Dk_*^2 + \partial_U F(0;0) - \lambda\right)]|_{\lambda=i\omega_*} \neq 0. \tag{6}$$

By Hypothesis 2.5 we may solve (4) for λ with k close to k_* for $\lambda = \lambda_*(k)$, with $\lambda_*(k_*) = i\omega_*$. Since $\det(-Dk^2 + \partial_U F(0;0) - \lambda) \neq 0$ for $\operatorname{Re}\lambda > 0$ and $k \in \mathbb{R}$, we necessarily have $\operatorname{Re}\partial_k\lambda_*|_{k=k_*} = 0$. We assume that the second derivative does not vanish.

HYPOTHESIS 2.6. *[Quadratic tangency]* We assume

$$\operatorname{Re}\frac{d^2\lambda_*(k)}{dk^2}|_{k=k_*} < 0, \tag{7}$$

where $\lambda_*(k)$ was defined above as the spectral curve close to the imaginary axis.

In case of a Turing-Hopf instability, there is another nontrivial quantity arising at the linear level, namely the derivative of the imaginary part of the dispersion relation $\operatorname{Im}\frac{d\lambda_*(k)}{dk}|_{k=k_*} =: c_g$.

HYPOTHESIS 2.7. *[Nonzero group velocity]* In case of a Turing-Hopf instability (TH), we assume $c_g \neq 0$.

So far we have collected the main assumptions on the possible linearizations of the reaction-diffusion system at criticality. The remainder of the section collects assumptions on typical unfoldings: we assume that the spectrum crosses the imaginary axis transversely when we vary the parameter μ.

In case of a Turing, Hopf, or Turing-Hopf instability, $\partial_U F(0;0)$ is invertible and the equilibrium $U \equiv 0$ continues to a unique family of spatially homogeneous equilibria $U(\mu)$ by the implicit function theorem. Shifting $\tilde{U} = U - U(\mu)$, we may assume that $F(0;\mu) = 0$. Consider then the parameterized version of (4)

$$d(k,\lambda;\mu) = \det\left(-Dk^2 + \partial_U F(0;\mu) - \lambda\right) = 0. \tag{8}$$

The implicit function theorem gives us a unique family of curves $\lambda_*(k;\mu)$ which we assume to cross the imaginary axis upon varying μ.

HYPOTHESIS 2.8. *[Transverse crossing]* In cases [T], [H], and [TH], we assume that $F(0;\mu) = 0$ for μ close to zero and $\partial_\mu d(k_*, i\omega_*;\mu)|_{\mu=0} \neq 0$.

If we denote by $\mathrm{Re}\,\lambda_{\mathrm{max}}(\mu)$ the unique maximum of the real part of $\lambda(k;\mu)$, then Hypothesis 2.8 ensures $\mathrm{d}\lambda_{\mathrm{max}}/\mathrm{d}\mu \neq 0$ at $\mu = 0$.

In case of a stationary homogeneous instability, the equilibrium $U \equiv 0$ can typically not be uniquely continued for $\mu \neq 0$. We then assume that varying the parameter actually destroys the trivial equilibrium. Let U^* span the kernel of the adjoint linearization of the kinetics with respect to some chosen scalar product $(\cdot,\cdot)$.

HYPOTHESIS 2.9. We assume $(U^*, \partial_\mu F(0;\mu)|_{\mu=0}) \neq 0$.

DEFINITION 2.10. *[Classification of linear instabilities]* Under Hypotheses 2.2, 2.5, 2.6, and 2.7 in case (TH), we say that the instability is *linearly generically unfolded* if

- Hypothesis 2.8 is satisfied in cases (T), (H), and (TH).
- Hypothesis 2.9 is satisfied in case (O).

We remark here that the set of matrices D and functions $F(U;\mu)$ with a linearly generic unfolding of a marginal stability according to Definition 2.10 is open and dense within the set of D and F which undergo a marginal instability, Hypothesis 2.2. Figure 2 with a sketch of the dispersion relations summarizes the discussion of this section.

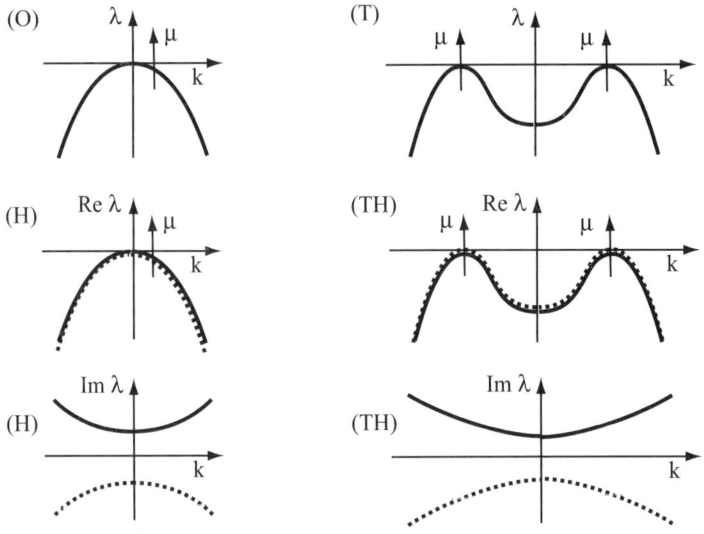

FIGURE 2. *Dispersion relations between temporal eigenvalue λ and wave number k in cases (O) -left- and (T) -right- in the top row. Real and imaginary parts of λ in cases (H) -left- and (TH) -right- at the bottom. In the oscillatory cases, the complex conjugate critical curve is shown as a dashed line.*

3. Stationary bifurcations and spatial dynamics

We consider stationary instabilities (O) or (T). We first rephrase the stationary reaction-diffusion system as a dynamical system in spatial time x and reinterpret the linear instabilities in terms of the linearization of this dynamical system; Section 3.1. We then state the center manifold reduction theorem, Section 3.2, which leads to a low-dimensional ODE. Using normal form theory as in Section 3.3, we find universal bifurcation equations. We analyze these bifurcation equations separately in case (O), Section 3.4, and in case (T) of a Turing instability, Section 3.5.

3.1. Spatial dynamics. We look for time-independent solutions $U(t,x) \equiv U(x)$, which are close to the trivial equilibrium $U \equiv 0$ in $BC^0_{\mathrm{unif}}(\mathbb{R},\mathbb{R}^N)$. The solution $U(x)$ then solves the ordinary differential equation $DU_{xx} + F(U;\mu) = 0$. We rewrite this equation as a first-order system

$$\begin{aligned} u_x &= w \\ w_x &= -D^{-1}F(u;\mu), \end{aligned} \tag{9}$$

or, in more compact notation,

$$\underline{u}_x = \mathcal{A}\underline{u} + \mathcal{F}(\underline{u};\mu). \tag{10}$$

Here, $\underline{u} = (u,w)^T$ belongs to the phase space $Y := \mathbb{R}^{2N}$, the linear part is defined by $\mathcal{A}(u,w)^T = (w, -D^{-1}\partial_U F(0;0)u)^T$, and the nonlinearity is given by $\mathcal{F}((u,w)^T;\mu) = (0, -D^{-1}(F(u;\mu) - \partial_U F(0;0)u))^T$. The linearization of (10) is $\underline{u}_x = \mathcal{A}\underline{u}$.

The reflectional symmetry $(x \mapsto -x)$ of the original partial differential equation (2) translates into reversibility of the ordinary differential equation (10): if $\underline{u}(x)$ is a solution, $R\underline{u}(-x)$ is a solution, where $R(u,w)^T = (u,-w)^T$.

We decompose the spectrum $\operatorname{spec}\mathcal{A}$ according to

$$\operatorname{spec}\mathcal{A} = \Sigma^{\mathrm{s}} \cup \Sigma^{\mathrm{c}} \cup \Sigma^{\mathrm{u}}$$

where $\Sigma^{\mathrm{s}} \subset \mathbb{C}_- = \{\lambda|\ \operatorname{Re}\lambda < 0\}$, $\Sigma^{\mathrm{u}} \subset \mathbb{C}_+ = \{\lambda|\ \operatorname{Re}\lambda > 0\}$, and $\Sigma^{\mathrm{c}} \subset \mathrm{i}\mathbb{R}$. The corresponding generalized eigenspaces are denoted by E^{s}, E^{c}, and E^{u}, respectively. By reversibility, $E^{\mathrm{s}} = RE^{\mathrm{u}}$ and $E^{\mathrm{c}} = RE^{\mathrm{c}}$. In consequence, E^{c} is even-dimensional. Let P^{s}, P^{c}, and P^{u} denote the spectral projections on E^{s}, E^{c}, and E^{u}, respectively, along their relative spectral complement.

LEMMA 2.11. *Assume Hypotheses 2.2, 2.5, and 2.6. In case (O),* $\dim E^{\mathrm{c}} = 2$ *with eigenvalue* $\nu = 0$ *geometrically simple and algebraically double. We may choose coordinates such that*

$$\mathcal{A}|_{E^{\mathrm{c}}} = \begin{pmatrix} 0 & 1 \\ 0 & 0 \end{pmatrix}, \quad R|_{E^{\mathrm{c}}} = \begin{pmatrix} 1 & 0 \\ 0 & -1 \end{pmatrix}.$$

In case (T), $\dim E^{\mathrm{c}} = 4$ *with eigenvalue* $\nu = \mathrm{i}k_*$ *geometrically simple and algebraically double. We may choose (complex) coordinates for* $\underline{u}_c \in E^{\mathrm{c}}$ *such that* $\mathcal{A}$*,* R*, and complex conjugation* τ *act according to*

$$\mathcal{A}|_{E^{\mathrm{c}}} = \begin{pmatrix} \mathrm{i}k_* & 1 & 0 & 0 \\ 0 & \mathrm{i}k_* & 0 & 0 \\ 0 & 0 & -\mathrm{i}k_* & 1 \\ 0 & 0 & 0 & -\mathrm{i}k_* \end{pmatrix}, \quad R|_{E^{\mathrm{c}}} = \begin{pmatrix} 0 & 0 & 1 & 0 \\ 0 & 0 & 0 & -1 \\ 1 & 0 & 0 & 0 \\ 0 & -1 & 0 & 0 \end{pmatrix},$$

and

$$\tau|_{E^{\mathrm{c}}} = \begin{pmatrix} 0 & 0 & 1 & 0 \\ 0 & 0 & 0 & 1 \\ 1 & 0 & 0 & 0 \\ 0 & 1 & 0 & 0 \end{pmatrix}.$$

Proof. If $\nu = \mathrm{i}k \in \operatorname{spec}\mathcal{A}$, we have $\det(-Dk^2 + \partial_U F(0;0)) = 0$. In addition, if $\underline{u} = (u,w)^T \in \operatorname{Ker}(\mathcal{A} - \mathrm{i}k)$ is in the kernel, then $w = \mathrm{i}ku$ and $u \in \operatorname{Ker}(-Dk^2 + \partial_U F(0;0))$. Hypothesis 2.5 implies that the geometric multiplicity of 0 (or $\mathrm{i}k_*$) is one in case (O) (or (T)). The algebraic multiplicity coincides with the order of tangency of the dispersion relation $\lambda_*(k)$ to the imaginary axis:

$$\begin{aligned}\det(\mathcal{A} - \nu) &= \det\begin{pmatrix} -\nu & 1 \\ -D^{-1}\partial_U F(0;0) & -\nu \end{pmatrix} = \det\left(\nu^2 + D^{-1}\partial_U F(0;0)\right) \\ &= \left(\det D^{-1}\right)\partial_\nu^2\left(\nu^2 D + \partial_U F(0;0)\right) \neq 0,\end{aligned}$$

in $\nu = 0$ or $\nu = \mathrm{i}k_*$, respectively, from Hypothesis 2.6.

It remains to choose particular coordinates to obtain the representation as claimed. First, since eigenvectors are of the form $(u, \nu u)^T$, the action of R on the eigenspace automatically takes the desired form. Next, in case (O), we choose the principal vector $(0, U_0)^T$ with $(U_0, 0)^T$ being the unique eigenvector, such that $R(0, U_0)^T = -(0, U_0)^T$. Parameterizing $\underline{u}_c = A(U_0, 0)^T + B(0, U_0)^T$ with $(A, B) \in \mathbb{R}^2$ gives the desired action of $\mathcal{A}$ and R.

In case (T), let $U_0 \in \mathbb{R}^N$ span the kernel of $k_*^2 - D^{-1}\partial_U F(0;0)$. Then the vector $(U_0, \pm\mathrm{i}k_* U_0)^T$ spans the kernel of $\mathcal{A} \mp \mathrm{i}k_*$. A principal vector $(u_\pm, w_\pm)^T$ to the eigenvalue $\pm\mathrm{i}k_*$ solves

$$\begin{aligned}-D^{-1}\partial_U F(0;0)u_\pm + k_*^2 u_\pm &= \pm 2\mathrm{i}k_* U_0 \\ w_\pm &= U_0 \pm \mathrm{i}k_* u_\pm.\end{aligned}$$

Note that $\operatorname{Re} u_\pm \in \operatorname{Ker}(k_*^2 - D^{-1}\partial_U F(0;0))$ such that we may choose $u_\pm$ purely imaginary with $u_- = -u_+ = \overline{u}_+$. Then $(u_+, w_+)^T = -R(u_-, w_-)^T = \overline{(u_-, w_-)}^T$. Parameterizing the center space by

$$\underline{u}_c = A(U_0, \mathrm{i}k_* U_0)^T + B(u_+, w_+)^T + \text{c.c.}$$

gives the desired representation of $\mathcal{A}$, R, and τ on $A, B, \overline{A}, \overline{B}$. ∎

The choice of coordinates we use is the same as in [**IP93**]; see [**MSW94**] for a slightly different choice.

REMARK 2.12. *Lemma 2.11 illustrates the relation between the dispersion relation and the eigenvalues in spatial dynamics in a particular example. In general, we find a curve $\lambda_j(k)$ near $\lambda = 0$ to each eigenvector U_j in the kernel of $\mathcal{L}_0$. The curves are parameterized by k close to k_j, where $\mathrm{i}k_j$ is a purely imaginary eigenvalue in spatial dynamics. The number of curves equals the dimension of the kernel and, at the same time, gives the sum of the geometric multiplicities of eigenvalues $\mathrm{i}k_j$ in spatial dynamics on the imaginary axis; see Lemma 2.1. The order of tangency t_j of the curves $\lambda_j(k)$ with the imaginary axis gives the algebraic multiplicity $a_j = t_j + 1$ of the eigenvalue $\mathrm{i}k_j$ in spatial dynamics. Also, particular bases of the center eigenspace can be constructed from the dispersion relation. The eigenvectors can be computed directly from the kernel of $-Dk_*^2 + \partial_U F(0;0)$. The principal vectors can be obtained from the dispersion relation using derivatives with respect to the wave-vector k: if $U_0(k)$ denotes the vector in the kernel of $-Dk^2 + \partial_U F(0;0) - \lambda(k)$, with $\lambda(k)$ from Hypothesis 2.6, then $-\mathrm{i}U_0'(k) = u_+$ is the first component of the principal vector to the eigenvalue $\mathrm{i}k_*$.*

A similar observation holds in case of an oscillatory instability; see Lemma 2.30.

3.2. Center manifolds. We show that small bounded solutions to (9) lie on a locally invariant, low-dimensional manifold. Recall that a *local flow* on a manifold $\mathcal{M}$ is a map $\Phi : \mathcal{U} \subset \mathbb{R} \times \mathcal{M} \to \mathcal{M}$, where $\mathcal{U}$ is an open neighborhood of $\{0\} \times \mathcal{M}$, which satisfies the flow properties $\Phi(0, u) = u$ and, whenever $\{(s, u), (t, \Phi(s, u))\} \subset \mathcal{U}$, we have $(t + s, u) \in \mathcal{U}$ and $\Phi(t, \Phi(s, u)) = \Phi(t + s, u)$. A *trajectory* through a point u_0 of the local flow is the set $\{\Phi(t, u_0);\ t \in \mathbb{R} \text{ with } (t, u_0) \in \mathcal{U}\}$. Given a differential equation, here the equation (9), we first introduce the notion of a center manifold.

DEFINITION 2.13. *[C^m-Center manifold]* A C^m-center manifold $\mathcal{W}^c \subset Y$ close to an equilibrium $\underline{u} = 0$ of a differential equation is an open C^m-manifold together with a local flow Φ^c on $\mathcal{W}^c$ such that

- trajectories of Φ^c are solutions of the underlying differential equation;
- there is a $\delta > 0$ such that for any solution $\underline{u}(x)$, $x \in \mathbb{R}$, with $|\underline{u}(x)|_Y < \delta$ for all $x \in \mathbb{R}$, we have $\underline{u}(0) \in \mathcal{W}^c$.

THEOREM 2.14. [Center manifolds] *Assume (9) undergoes a stationary instability and let E^c denote the center eigenspace of $\mathcal{A}$.*

Then for any $0 < m < \infty$, there is $\delta > 0$ such that for all $|\mu| < \delta$ there exists a C^m-center manifold $\mathcal{W}^c \subset Y$. The dependence of $\mathcal{W}^c$ as a manifold on μ is C^m. For $\mu = 0$, it is tangent to E^c in $\underline{u} = 0$. Moreover, $R\mathcal{W}^c = \mathcal{W}^c$ and $R\Phi^c(x, \underline{u}) = \Phi^c(-x, R\underline{u})$, whenever one of both is defined.

There are various proofs of this theorem in the literature. We refer to [**Shu87**] for a geometric proof via graph transform and to [**Van89, VI91**] for Perron's method, using variation-of-constant formulas. We exploit both methods in Chapters 3 and 4, to construct invariant manifolds in a different, nonautonomous setting.

Since $\mathcal{W}^c$ is tangent to E^c at $\mu = 0$, we may write

$$\mathcal{W}^c = \{(\underline{u}^c, \psi(\underline{u}^c; \mu))\,;\, |\underline{u}^c|, |\mu| \leq \delta'\}$$

with δ' small and $\psi \in C^m(E^c \times \mathbb{R}^p, E^s \oplus E^u)$. The flow on $\mathcal{W}^c$ is generated by the vector field of (9), restricted to $\mathcal{W}^c$:

$$\underline{u}^c_x = \mathcal{A}^c\underline{u}^c + P^c\mathcal{F}\left(\underline{u}^c + \psi(\underline{u}^c; \mu); \mu\right) =: g^c(\underline{u}^c; \mu), \tag{11}$$

where $\mathcal{A}^c := \mathcal{A}|_{E^c}$, and the hyperbolic component of $\underline{u}$ is given as

$$(P^s + P^u)\underline{u} = \psi(\underline{u}^c; \mu). \tag{12}$$

We call (11) the *reduced equation.* Taylor jets can be computed recursively from the invariance condition (12), which gives

$$\partial_u\psi(\underline{u}^c; \mu)g^c(\underline{u}^c; \mu) = (P^s + P^u)\left(\mathcal{A}\psi(\underline{u}^c) + \mathcal{F}(\underline{u}^c + \psi(\underline{u}^c; \mu); \mu)\right). \tag{13}$$

At each recursion step, we compute the ℓ'th order Taylor monomials of ψ, making use of the tangency $\psi(0; 0) = 0$, $\partial_u\psi(0; 0) = 0$.

In case of a stationary homogeneous instability, expanding the reduced vector field leads to the following proposition.

PROPOSITION 2.15. *In case of a linearly generically unfolded instability of type (O), the reduced vector field reads in appropriate coordinates*

$$\begin{aligned} A_x &= B \\ B_x &= (\gamma_0, \mu) + (\gamma_1, \mu)A + \gamma_2 A^2 + \gamma_3 A^3 + \mathcal{R}(A, B; \mu) \end{aligned} \tag{14}$$

with amplitudes $A, B \in \mathbb{R}$. *Again,* $(\cdot,\cdot)$ *denotes a fixed scalar product in* $\mathbb{R}^p$. *The equation is invariant under reversibility* $(x \mapsto -x, B \mapsto -B)$*; in particular,* $\mathcal{R}(A, -B; \mu) = \mathcal{R}(A, B; \mu)$. *The coefficients* $\gamma_0, \gamma_1 \in \mathbb{R}^p$ *and* $\gamma_2, \gamma_3 \in \mathbb{R}$ *can be computed from (13). Generic unfolding implies* $\gamma_0 \neq 0$. *The corrections satisfy the estimate*

$$\mathcal{R}(A, B; \mu) = \mathrm{O}\left(|\mu|^2 + |\mu|A^2 + |AB^2| + |B|^3 + (|A| + |B|)^4\right).$$

Small solutions of (14) yield small solutions of the full system (9) via the parameterization $\underline{u}^{\mathrm{c}} = A(U_0, 0)^T + B(0, U_0)^T + \mathrm{O}((|\mu| + |A| + |B|)^3)(0, U_0)^T$ *and lifting to the manifold* $\underline{u} = \underline{u}^{\mathrm{c}} + \psi(\underline{u}^{\mathrm{c}}; \mu)$. *Here,* $U_0 \neq 0$ *spans the kernel of* $\partial_U F(0; 0)$. *The original variable* u *from (9) is recovered from* $u = AU_0 + O(|\mu| + |A|^2)$.

Proof. First write $\underline{u}^{\mathrm{c}} = A(U_0, 0)^T + \tilde{B}(0, U_0)^T$, in the coordinates of Lemma 2.11. We obtain

$$A_x = \tilde{B} + \tilde{\mathcal{R}}_1(A, \tilde{B}; \mu), \quad \tilde{B}_x = \tilde{\mathcal{R}}_2(A, \tilde{B}; \mu)$$

and $\tilde{\mathcal{R}}_j(A, \tilde{B}; \mu) = \mathrm{O}(|\mu| + (|\mu| + |A| + |\tilde{B}|)^2)$, $j = 1, 2$. From (13), we compute the quadratic terms of the vector field by evaluating the nonlinearity on E^{c} and then projecting with P^{c}. We obtain

$$\begin{aligned} \tilde{\mathcal{R}}_1(A, \tilde{B}; \mu) &= \mathrm{O}\left(|\mu|(|A| + |\tilde{B}|) + |\mu|^2 + (|A| + |\tilde{B}|)^2\right), \\ \tilde{\mathcal{R}}_2(A, \tilde{B}; \mu) &= \gamma_2 A^2 + \mathrm{O}\left(|\mu| + (|A| + |\tilde{B}|)^3\right). \end{aligned}$$

We next change coordinates to $B = \tilde{B} + \tilde{\mathcal{R}}_1(A, \tilde{B}; \mu)$. Expanding the terms in the equation for B_x then proves the proposition. ∎

In case of a Turing instability, the analogue of Proposition 2.15 requires some preparation. We concentrate on the linear part of the equation, first.

PROPOSITION 2.16. *Consider a linearly generically unfolded Turing instability (T). After a linear change of coordinates, smoothly depending on* μ*, and* μ*-close to the coordinates of Lemma 2.11 on* E^{c}*, we obtain the reduced equation for the complex amplitudes* A, B

$$\begin{aligned} A_x &= \mathrm{i}k_* A + B + \mathrm{i}k_1(\mu)A + \mathrm{O}\left((|A| + |B|)^2\right) \\ B_x &= \gamma_1(\mu)A + \mathrm{i}k_* B + \mathrm{i}k_1(\mu)B + \mathrm{O}\left((|A| + |B|)^2\right) \end{aligned} \tag{15}$$

with γ_1 *and* k_1 *smooth, real functions of* μ, $\gamma_1(0) = k_1(0) = 0$ *and* $\gamma_1'(0) \neq 0$.

Proof. Note that for $\mu = 0$, the form of the equation is as described in Lemma 2.11. By means of a smooth linear change of coordinates, the μ-dependent linearization in $\underline{u}^{\mathrm{c}} = 0$ of the reduced equation (11) on the center manifold can be transformed to the above form, sometimes called the Arnol'd normal form of the versal deformation; see [**Arn83, ETBCI87**] and Section 3.3. We have to show that $\gamma_1'(0) \neq 0$. First observe $\gamma_1'(0) = -\partial_\mu[\det(\mathcal{A}(\mu) - \mathrm{i}k_*)]_{\mu=0}$, where

$$\mathcal{A}(\mu) = \begin{pmatrix} 0 & 1 \\ -D^{-1}\partial_U F(0; \mu) & 0 \end{pmatrix}.$$

Then $\det(\mathcal{A}(\mu) - \mathrm{i}k_*) = \det(-Dk_*^2 + \partial_U F(0; \mu))/\det D$. By Hypothesis 2.8, the derivative $\partial_\mu[-Dk_*^2 + \partial_U F(0; \mu)]$ does not vanish in $\mu = 0$, which proves the proposition. ∎

3.3. Normal forms. In this section, we review normal form theory as developed in [**ETBCI87**]. We start with a general differential equation in $\mathbb{R}^M$, M arbitrary, with equilibrium $u = 0$ and linearization $\mathcal{B}$ in $u = 0$. We will later apply the results to reduced equations on center manifolds, (11). Consider

$$u_x = \mathcal{B}u + \mathcal{G}(u;\mu) \in \mathbb{R}^M, \tag{16}$$

with $\operatorname{spec}\mathcal{B} \subset \mathrm{i}\mathbb{R}$ and $\mathcal{G}(u;\mu) = \mathrm{O}((|\mu| + |u|)|u|)$, we ask for its 'simplest' form after a polynomial change of coordinates

$$u = v + \Psi(v;\mu), \quad \Psi(v;\mu) = \mathrm{O}\left((|\mu| + |v|)|v|\right).$$

THEOREM 2.17. [Normal form [**ETBCI87**]] *For every $0 < m < \infty$, there exists a smooth change of coordinates $u = v + \Psi(v;\mu)$, polynomial in v of degree m, such that in the new variable v, equation (16) takes the normal form*

$$v_x = \mathcal{B}v + \mathcal{N}(v;\mu) + \mathcal{R}(v;\mu), \tag{17}$$

with $\mathcal{R}(v;\mu) = \mathrm{O}(|v|^{m+1})$ a small remainder. The normal form part $\mathcal{N}$ is a polynomial of degree m in v, with μ-dependent, smooth coefficients, and it commutes with the flow to the adjoint linearized equation:

$$\mathrm{e}^{\mathcal{B}^*\tau}\mathcal{N}(v;\mu) = \mathcal{N}(\mathrm{e}^{\mathcal{B}^*\tau}v;\mu).$$

The adjoint $\mathcal{B}^*$ is taken with respect to some chosen scalar product.

If (16) is reversible, then (17) is reversible. If (16) is equivariant with respect to a compact subgroup of the orthogonal group $\mathcal{O}(M)$, then (17) is equivariant, too: for all $\gamma \in \mathcal{O}(M)$, we have

$$\mathcal{N}(\gamma v;\mu) = \gamma\mathcal{N}(v;\mu), \quad \mathcal{R}(\gamma v;\mu) = \gamma\mathcal{R}(v;\mu)$$

whenever the relations hold for $\mathcal{B}$ and $\mathcal{G}$.

We outline the proof of the theorem since we will have to modify the construction in Chapter 3 to find a time-dependent normal form for nonautonomous systems.

The proof uses induction on m. We suppress μ-dependence throughout. Suppose we have already transformed the equation into normal form up to order $m-1$:

$$u_x = \mathcal{B}u + \sum_{j=1}^{m-1} \mathcal{N}_j(u) + \tilde{\mathcal{N}}_m(u) + \mathcal{R}(u),$$

with $\mathcal{N}_j$ and $\tilde{\mathcal{N}}_m$ being homogeneous polynomials in u of degree j and m, $\mathcal{R}(u) = \mathrm{O}(|u|^{m+1})$, and assume that $\mathcal{N}_j$, $1 \le j \le m-1$, are in normal form. We set $u = v + \psi(v)$ with ψ a homogeneous polynomial of degree m in v. The equation for v then becomes

$$v_x = \mathcal{B}v + \sum_{j=1}^{m-1} \mathcal{N}_j(v) + \tilde{\mathcal{N}}_m(v) + \mathcal{B}\psi(v) \quad \partial_v\psi(v)\mathcal{B}v + \tilde{\mathcal{R}}(v),$$

where $\tilde{\mathcal{R}}(v) = \mathrm{O}(|v|^{m+1})$. The new equation is in normal form up to order m, if, and only if,

$$N_m(v) := \tilde{\mathcal{N}}_m(v) + \mathcal{B}\psi(v) - \partial_v\psi(v)\mathcal{B}v$$

is in normal form.

We view $\mathcal{B}\psi(v) - \partial_v\psi(v)\mathcal{B}v =: (\mathrm{ad}_m\mathcal{B}\psi)(v)$ as a linear operator on the space $\mathcal{P}_m$ of homogeneous polynomials of degree m. We decompose $\mathcal{P}_m = \mathcal{S}_m \oplus \mathrm{Rg}\,(\mathrm{ad}_m\mathcal{B})$

with a suitable complement $\mathcal{S}_m$. By Fredholm's alternative, we may choose $\mathcal{S}_m = \mathrm{Ker}\,((\mathrm{ad}_m\mathcal{B})^*)$. The main observation in [**ETBCI87**] is that we may choose a scalar product in $\mathcal{P}_m$ such that $(\mathrm{ad}_m\mathcal{B})^* = \mathrm{ad}_m(\mathcal{B}^*)$. We may therefore choose $\psi \in \mathcal{P}_m$ such that $N_m \in \mathrm{Ker}\,(\mathrm{ad}_m(\mathcal{B}^*))$. Now the differential equation $\mathrm{ad}_m(\mathcal{B}^*)N_m = 0$ on $\mathcal{P}_m$ can be integrated to the equivariance condition $\mathrm{e}^{\mathcal{B}^*\tau}\mathcal{N}(v) - \mathcal{N}(\mathrm{e}^{\mathcal{B}^*\tau}v) = 0$. For $\mu \neq 0$, small, we solve the μ-dependent problem using Lyapunov-Schmidt reduction. The necessary decomposition for the linear part is given by $\mathcal{P}_m = \mathrm{Rg}\,(\mathrm{ad}_m\mathcal{B}) \oplus \mathrm{Ker}\,((\mathrm{ad}_m\mathcal{B})^*)$.

The particular case of a Turing instability has been considered in [**ETBCI87**].

PROPOSITION 2.18. [**ETBCI87**] *Consider a linearly generically unfolded Turing instability (T). Fix $0 < m < \infty$ arbitrary. Then, there exists a polynomial change of coordinates such that the reduced vector field for the complex amplitudes $A, B \in \mathbb{C}$ is given by*

$$
\begin{aligned}
A_x &= \mathrm{i}k_* A + B + \mathrm{i}AP_1 + \mathcal{R}_1 \\
B_x &= \gamma_1(\mu)A + \mathrm{i}k_* B + \mathrm{i}BP_1 + AP_2 + \mathcal{R}_2
\end{aligned}
\tag{18}
$$

where $P_j = P_j(I, J; \mu)$ are real polynomials in the real arguments $I = |A|^2$ and $J = \mathrm{i}(A\overline{B} - \overline{A}B)$, with smooth, μ-dependent coefficients and $P_j(0,0;0) = 0$, $P_2(0,0;\mu) = 0$. The linear unfolding parameter $\gamma_1(\mu)$ is a smooth, real-valued function with $\gamma_1'(0) \neq 0$. The remainder terms satisfy $\mathcal{R}_j = \mathrm{O}(|A| + |B|)^{m+1}$.

REMARK 2.19. *If we set $\tilde{B} = B + \mathrm{i}AP_1$, the system (18) transforms to*

$$
\begin{aligned}
A_x &= \mathrm{i}k_* A + \tilde{B} + \tilde{\mathcal{R}}_1 \\
\tilde{B}_x &= \gamma_1(\mu)A + \mathrm{i}k_* \tilde{B} + A\tilde{P}_2 + \tilde{\mathcal{R}}_2,
\end{aligned}
\tag{19}
$$

with $\tilde{\mathcal{R}}_j = \mathrm{O}((|A| + |\tilde{B}|)^{m+1})$. In a corotating frame, $\hat{A} = A\mathrm{e}^{-\mathrm{i}k_* x}$, $\hat{B} = \tilde{B}\mathrm{e}^{-\mathrm{i}k_* x}$, we obtain

$$
\begin{aligned}
\hat{A}_x &= \hat{B} + \hat{\mathcal{R}}_1 \\
\hat{B}_x &= \gamma_1(\mu)\hat{A} + \hat{A}\tilde{P}_2 + \hat{\mathcal{R}}_2.
\end{aligned}
\tag{20}
$$

Here, P_2 is, as before, a function of $\hat{I} = |\hat{A}|^2, \hat{J} = \mathrm{i}(\hat{A}\overline{\hat{B}} - \overline{\hat{A}}\hat{B})$. The remainder terms $\hat{\mathcal{R}}_j(A, B; x) = \mathrm{O}((|A| + |B|)^{m+1})$ now are explicitly depending on spatial time x, with period $2\pi/k_*$.

3.4. Fold and cusp. We analyze the reduced equation in case of a stationary, homogeneous instability. From Proposition 2.15, we have the reduced equation (14) for the real amplitudes A, B, with expansion

$$
\begin{aligned}
A_x &= B \\
B_x &= (\gamma_0, \mu) + (\gamma_1, \mu)A + \gamma_2 A^2 + \gamma_3 A^3 + \mathcal{R}(A, B; \mu).
\end{aligned}
$$

We first analyze the generic case of the fold.

HYPOTHESIS 2.20. *[Fold]* Consider a linearly generically unfolded homogeneous stationary instability (O). Assume that $\gamma_2 > 0$.

If $\gamma_2 < 0$, we replace U by $-U$ in the original reaction-diffusion system (2) and find $\gamma_2 > 0$ after the transformation.

PROPOSITION 2.21. *Assume Hypothesis 2.20 with one-dimensional parameter μ. Then for each μ small enough with $\mu\gamma_0 < 0$, there exists a solution of the reduced equation (14), $q(x;\mu) = (Q(x;\mu), Q'(x;\mu))$, smoothly depending on μ, with expansion*

$$Q(x;\mu) = \left|\frac{\mu\gamma_0}{\gamma_2}\right|^{1/2} \left(1 - 3\,\mathrm{sech}^2\left(|\gamma_0\gamma_2\mu/4|^{1/4}x\right)\right) + \mu Q_R(|\mu|^{1/4}x).$$

The correction Q_R and its derivatives are bounded. The solution is unique (up to translations in x) as a nonconstant pulse solution, that is, with asymptotics $Q(x;\mu) \to |\frac{\mu\gamma_0}{\gamma_2}|^{1/2} + \mathrm{O}(|\mu|)$.

The phase portrait of the reduced equation is depicted in Figure 3. In the original reaction-diffusion system, we find a pulse solution, that is, an exponentially localized, stationary pattern, with monotonically decreasing derivative, asymptotic to the stable background state for $x \to \pm\infty$.

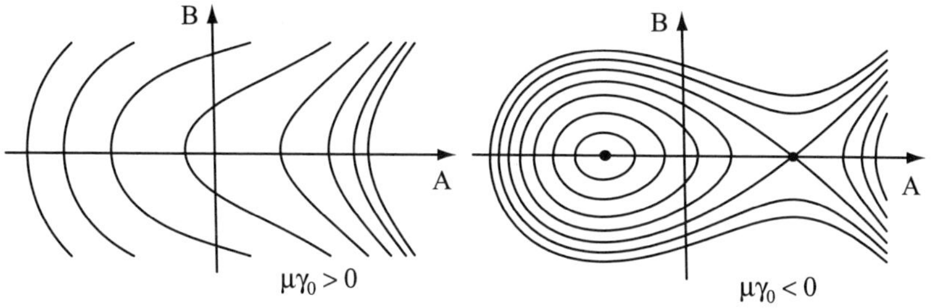

FIGURE 3. *The reduced flow in case of a fold in the sub- and supercritical parameter range. The homoclinic in the right picture is the particularly interesting localized pulse solution.*

Proof. We rescale $A = (-\frac{\mu\gamma_0}{\gamma_2})^{1/2}\tilde{A}$ and $x = (-\gamma_0\gamma_2\mu)^{-1/4}\tilde{x}$ to obtain

$$\tilde{A}_{\tilde{x}} =: \tilde{B}, \qquad \tilde{B}_{\tilde{x}} = -1 + \tilde{A}^2 + \tilde{\mathcal{R}} \tag{21}$$

where $\tilde{\mathcal{R}} = \mathrm{O}(|\mu|^{1/2})$. The full equation is reversible with respect to $(\tilde{x} \mapsto -\tilde{x}, \tilde{B} \mapsto -\tilde{B})$. The truncated equation, with $\tilde{\mathcal{R}}$ set to zero, possesses a unique homoclinic solution $\tilde{q}(\tilde{x}) = (\tilde{Q}, \tilde{Q}')(\tilde{x})$ with $\tilde{Q}(\tilde{x}) = 1 - 3\,\mathrm{sech}^2(\tilde{x}/\sqrt{2})$. The homoclinic is symmetric in $\tilde{x}$. We have to show persistence of the homoclinic for the full equation. First, the equilibrium $p_1 = (1,0)^T$ continues as a hyperbolic equilibrium to $p_1(\mu) = (1,0)^T + \mathrm{O}(|\mu|^{1/2})$ for (20). Denote by $\mathcal{W}^{\mathrm{u}}(\mu)$ its unstable manifold. On any compact part, $\mathcal{W}^{\mathrm{u}}(\mu)$ and $\mathcal{W}^{\mathrm{u}}(0)$ are $\mathrm{O}(|\mu|^{1/2})$-close in the C^1-topology. In particular, $\mathcal{W}^{\mathrm{u}}(\mu)$ intersects the reversibility line $\mathrm{Fix}\,R = \{(A,0); A \in \mathbb{R}\}$ transversely in a point $p_{\mathrm{s}}(\mu) = (-2+\mathrm{O}(|\mu|^{1/2}), 0)^T$. Next, note that stable and unstable manifolds of the equilibrium p_1 are conjugate to each other by the reflection R: $\mathcal{W}^{\mathrm{s}}(\mu) = R\mathcal{W}^{\mathrm{u}}(\mu)$, and therefore $p_{\mathrm{s}}(\mu) \in \mathcal{W}^{\mathrm{s}}(\mu)$ as well. We have found the unique μ-dependent homoclinic point as intersection of $\mathcal{W}^{\mathrm{s}}(\mu)$ and $\mathcal{W}^{\mathrm{u}}(\mu)$ in $\mathrm{Fix}\,R$. Transforming back to the original scaling proves the proposition. ∎

REMARK 2.22. *The complete set of small bounded solutions of the reduced equation (14) consists of the homoclinic, the two equilibria, and a family of symmetric periodic orbits filling the region inside the homoclinic. All of them persist due to reversibility. In fact, the phase portrait of the truncated equation can be read off from the level lines of the Hamiltonian* $H(\tilde{A}, \tilde{B}) = \frac{1}{2}\tilde{B}^2 + \tilde{A} - \frac{1}{3}\tilde{A}^3$. *However, the Hamiltonian structure of the truncated equation need not be preserved by the higher-order terms, since the kinetics in the original reaction-diffusion system are not assumed to be a gradient* $F(U) = \nabla V(U)$ *and possibly* $D \neq \mathrm{id}$. *We therefore exploited reversibility for persistence.*

For the other sign of μ, *there are no small bounded solutions, as can be seen from the truncated equation* $\tilde{A}'' = 1 + \tilde{A}^2$ *and a flow-box argument.*

There are no small heteroclinic orbits corresponding to standing interfaces in the reaction-diffusion system.

We address the case of a cusp in the kinetics F, next. This codimension-two bifurcation is of particular interest, since stable states may coexist.

HYPOTHESIS 2.23. *[Cusp]* Consider a linearly generically unfolded homogeneous stationary instability (O) with two-dimensional parameter $\mu = (\mu_1, \mu_2)$. Assume that $\gamma_2 = 0$ and $\gamma_3 = 1$ in (14). Furthermore, assume that $\gamma_0 = (1, 0)$ and $\gamma_1 = (0, 1)$.

The particular values of γ_0, γ_1, and γ_3 can be achieved by rescaling the independent variable U of the original reaction-diffusion system (2) and transforming the parameter μ by a local diffeomorphism close to $\mu = 0$.

The reduced, truncated equation with Hypothesis 2.23 reads

$$A' = B, \quad B' = \mu_1 + \mu_2 A + A^3.$$

The equation possesses three equilibria $A_- < A_0 < A_+$ within the cuspoidal region $\mu_2 < 0$, $\mu_1^2 < -\frac{4}{27}\mu_2^3$; see Chapter 1, Figure 2.

The sign convention for γ_3 actually implies that the unique equilibrium outside the closure of the cuspoidal region is stable. Analyzing the reduced flow in this region, we would find the equilibrium to be the unique, small bounded solution. We therefore focus on the cuspoidal region. Along the boundaries of the cusp, we find the fold as analyzed in Proposition 2.21. The standing pulses from Proposition 2.21 exist inside the cuspoidal region and are asymptotic to A_- or A_+. The interesting, new phenomenon here are heteroclinic orbits, which occur along a curve in parameter space. They mark the transition between the homoclinics asymptotic to A_- and A_+, see Figure 4.

PROPOSITION 2.24. *There is a curve* $\mu_1 = h(\mu_2)$, $\mu_2 < 0$ *inside the cuspoidal region, with* $h(\mu_2) = \mathrm{O}(\mu_2^2)$ *such that for parameter values* μ *on this curve, there exists a heteroclinic orbit of the reduced equation (14), unique up to translation and reflection in* x. *It is given to leading order by* $q(x; \mu) = (Q(x; \mu), Q'(x; \mu))$ *with expansion*

$$Q(x; \mu) = |\mu_2|^{1/2} \tanh\left(|\mu_2/2|^{1/2} x\right) + \mu_2 Q_R\left(|\mu_2|^{1/2} x\right),$$

with Q_R *and its derivatives bounded.*

In the full reaction-diffusion system, the solution $q(x; \mu)$ from the proposition represents a standing interface solution with expansion $Q(x; \mu)U_0 + \mathrm{O}(\mu_2)$, which marks the parameter region of coexistence between two stable equilibria.

Again, we have sketched the "complete" bifurcation diagram, Figure 4.

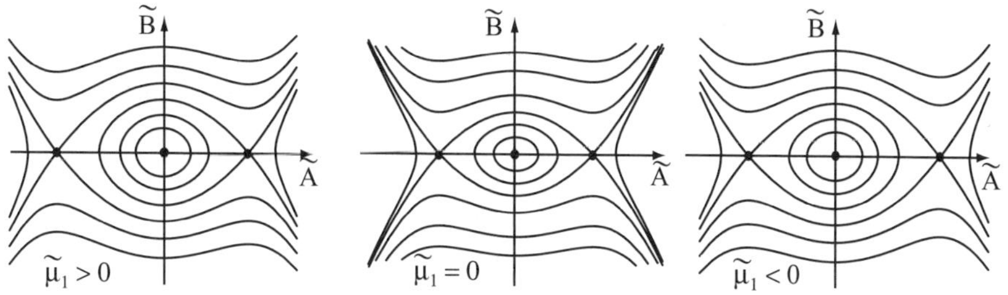

FIGURE 4. *The phase portrait of the reduced flow in the parameter region with coexistence of two stable equilibria, depending on $\tilde{\mu}_1$. The heteroclinic in the middle picture is the layer separating coexisting stable equilibria.*

Proof. The only region in parameter space which allows for coexistence of equilibria is found within the scaling $\mu_1 = \tilde{\mu}_1|\mu_2|^{3/2}$ with $\mu_2 < 0$ small and $|\tilde{\mu}_1|$ uniformly bounded in μ_2. We scale the amplitude $A = \tilde{A}|\mu_2|^{1/2}$, $x = \tilde{x}|\mu_2|^{-1/2}$, and obtain

$$\tilde{A}_{\tilde{x}} =: \tilde{B}, \qquad \tilde{B}_{\tilde{x}} = \tilde{\mu}_1 - \tilde{A} + \tilde{A}^3 + \tilde{\mathcal{R}} \tag{22}$$

with $\tilde{\mathcal{R}} = \mathrm{O}(|\mu_2|^{1/2})$. The truncated equation ($\tilde{\mathcal{R}} = 0$) possesses for $\tilde{\mu}_1 = 0$ a unique heteroclinic $\tilde{q}(\tilde{x})$ connecting $(\pm 1, 0)^T$, up to translation and reversibility. It is given explicitly through $\tilde{A}(\tilde{x}) = \tanh(\tilde{x}/\sqrt{2})$. It remains to show that this heteroclinic persists for the full equation and parameter values on a curve $\tilde{\mu_1} = \tilde{h}(|\mu_2|^{1/2})$, $\mu_2 < 0$.

Denote by $p_-(\tilde{\mu}_1), p_+(\tilde{\mu}_1)$ the equilibria in the truncated equation with $p_\pm(0) = (\pm 1, 0)$ and by $\mathcal{W}^{\mathrm{u}}_-(\tilde{\mu}_1)$ and $\mathcal{W}^{\mathrm{s}}_+(\tilde{\mu}_1)$ their unstable and stable manifolds, respectively — which depend smoothly on $\tilde{\mu}_1$. Clearly,

$$\tilde{q}(0) \in \bigcup_{\tilde{\mu}_1} \mathcal{W}^{\mathrm{u}}_-(\tilde{\mu}_1) \cap \bigcup_{\tilde{\mu}_1} \mathcal{W}^{\mathrm{s}}_+(\tilde{\mu}_1)$$

We claim that the intersection is transverse, that is,

$$\mathcal{T}_{\tilde{q}(0)}\left(\bigcup_{\tilde{\mu}_1} \mathcal{W}^{\mathrm{u}}_-(\tilde{\mu}_1)\right) + \mathcal{T}_{\tilde{q}(0)}\left(\bigcup_{\tilde{\mu}_1} \mathcal{W}^{\mathrm{s}}_+(\tilde{\mu}_1)\right) = \mathbb{R}^2 \times \mathbb{R}$$

in the extended phase space $(A, B; \tilde{\mu}_1) \in \mathbb{R}^2 \times \mathbb{R}$; see Figure 5. As a transverse

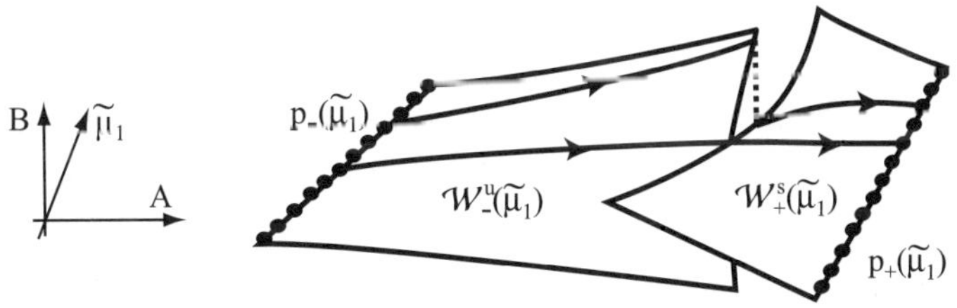

FIGURE 5. *Transverse intersection of stable and unstable manifolds $\bigcup_{\tilde{\mu}_1} \mathcal{W}^{\mathrm{u}}_-(\tilde{\mu}_1)$ and $\bigcup_{\tilde{\mu}_1} \mathcal{W}^{\mathrm{s}}_+(\tilde{\mu}_1)$ in the extended phase space.*

intersection, the heteroclinic would then persist, when switching on the μ_2-small perturbation $\mathcal{R}$, for nearby values $\tilde{\mu}_1 = \mathrm{O}(|\mu_2|^{1/2})$.

Transversality can be checked through a Melnikov-type computation. We write $g(A, B) = (\tilde{B}, \tilde{\mu}_1 - \tilde{A} + A^3)$. Let $\psi(\tilde{x})$ be a bounded solution of the adjoint variational equation

$$\psi_{\tilde{x}}(\tilde{x}) = -g'(\tilde{q}(\tilde{x}))\psi(\tilde{x}). \tag{23}$$

Since the intersection of stable and unstable manifolds is only one-dimensional, the nontrivial solution $\psi(x)$ is unique up to scalar multiples [**Lin90**]. Transverse intersection of stable and unstable manifolds in the extended phase space is equivalent to the Melnikov condition

$$M = \int_{\mathbb{R}} (\psi(\xi), \partial_{\tilde{\mu}_1} g\,(\tilde{q}(\xi)))\,\mathrm{d}\xi \neq 0 \tag{24}$$

in $\tilde{\mu}_1 = 0$; see [**GH, Lin90**]. The integral can be interpreted as the Lyapunov-Schmidt projection of the perturbation on the kernel of the linearization [**CH**]

We have [**Lin90**]

$$\psi(\tilde{x}) \perp \left(\mathcal{T}_{\tilde{q}(\tilde{x})}\mathcal{W}^{\mathrm{u}}_{-}(0) + \mathcal{T}_{\tilde{q}(\tilde{x})}\mathcal{W}^{\mathrm{s}}_{+}(0)\right). \tag{25}$$

Since $\tilde{q}'(\tilde{x}) > 0$, we have $\psi(\tilde{x}) = (\psi_1(\tilde{x}), \psi_2(\tilde{x}))$ with $\psi_2(\tilde{x}) > 0$ for all $\tilde{x}$. Also $\partial_{\tilde{\mu}_1} g(\tilde{q}(\tilde{x})) = (0,1)^T$ and therefore

$$\int_{\mathbb{R}} (\psi(\xi), \partial_{\tilde{\mu}_1} g(\tilde{q}(\xi)))\,\mathrm{d}\xi > 0,$$

which proves $M > 0$ and the proposition. ∎

REMARK 2.25. *As mentioned in Remark 2.22, the complete picture of small bounded solutions to the reduced equation (14) can be obtained from the truncated equation (22). In addition to layers, that is, heteroclinic orbits, we find homoclinic solutions within the cuspoidal region of coexistence of homogeneous equilibria* $|\mu_1| < \frac{2}{3\sqrt{3}}|\mu_2|^{3/2} + \mathrm{O}(\mu_2^2)$, $\mu_2 < 0$. *The region in phase space inside the homoclinic solution is again filled with periodic trajectories. All these solutions are symmetric in x and persist as mentioned in Remark 2.22.*

REMARK 2.26. *Instead of restricting to stationary solutions of the reaction-diffusion system (2), we could look more generally for travelling waves* $U(t, x) = Q(x - ct)$, *which solve*

$$u_\xi = w, \qquad w_\xi = -D^{-1}(F(u;\mu) + cw)$$

with $\xi = x - ct$ and μ, c small parameters.

Going through the previous reduction steps, we arrive at a reduced equation

$$A_\xi = B, \qquad B_\xi = (\gamma_0, \mu) + (\gamma_1, \mu)A + \gamma_2 A^2 + \gamma_3 A^3 - \bar{\gamma} c B + \mathcal{R}(A, B; \mu, c).$$

In analogy to (14), we have reversibility $(\xi \mapsto -\xi, B \mapsto -B, c \mapsto -c)$; *in particular,* $\mathcal{R}(A, -B; \mu, -c) = \mathcal{R}(A, B; \mu, c)$. *The remainder satisfies the estimate*

$$\mathcal{R}(A, B; \mu, c) = \mathrm{O}((|\mu| + |c|)^2 + (|\mu| + |c|)A^2 + |AB^2| + (|A| + |B|)^4).$$

The coefficient $\bar{\gamma}$ is always positive.

We find heteroclinic orbits of saddle-sink type, corresponding to fronts of a PDE-stable equilibrium invading a PDE-unstable equilibrium. These fronts and their stability have been analyzed in [**KR96, RK98**] *in the similar case of a transcritical bifurcation. The minimal wave speed c scales with $c = \tilde{c}|\mu_2|^{1/2}$. The layers* *found in $c = 0$ continue to saddle-saddle heteroclinics with a distinguished wave speed $c = \tilde{c}|\mu_2|^{1/2}$ inside the cuspoidal region. The heteroclinics represent a stable state invading another stable state.*

3.5. Turing instabilities. We first study the truncated normal form equation. We focus on the case of a weakly subcritical instability, which allows for coexistence of stable Turing patterns and the stable homogeneous equilibrium. We find standing interfaces for certain parameter values and argue that, typically, standing interfaces exist for open sets of parameter values.

A different kind of competition is observed in the supercritical regime, when the dynamics are governed by the Ginzburg-Landau equation $A_t = A_{xx} + A - A|A|^2$. Interaction between stationary periodic patterns $\sqrt{1-k_\pm^2}\mathrm{e}^{\mathrm{i}k_\pm x}$ on the left and on the right half line may lead to front solutions [**EG93**] or diffusive repair [**BK92, CE92, GM98**].

3.5.1. *Fifth order normal form.* From Remark 2.19, we have the expanded normal form, when redefining $\tilde{B}$ appropriately,

$$
\begin{aligned}
\tilde{A}_{\tilde{x}} &= \tilde{B} + \mathrm{i}k_*\tilde{A} \\
\tilde{B}_{\tilde{x}} &= \mathrm{i}k_*\tilde{B} + \gamma_1(\mu)\tilde{A} + \gamma_2(\mu)\tilde{A}|\tilde{A}|^2 \\
&\quad + \gamma_3(\mu)\mathrm{i}\tilde{A}(\overline{\tilde{A}}\tilde{B} - \tilde{A}\overline{\tilde{B}}) + \gamma_4(\mu)\tilde{A}|\tilde{A}|^4 + \tilde{\mathcal{R}}(\tilde{A},\tilde{B};\mu),
\end{aligned}
\tag{26}
$$

with smooth, μ-dependent coefficients $\gamma_j(\mu)$. The remainder satisfies

$$
\tilde{\mathcal{R}} = \mathrm{O}\left(|\tilde{A}\tilde{B}^2| + \sum_{j=1}^{5}|\tilde{B}^j\tilde{A}^{5-j}| + (|\tilde{A}| + |\tilde{B}|)^7\right).
$$

3.5.2. *Weakly subcritical instabilities.* We assume $\gamma_2(0) = 0$, an additional degeneracy in the equation. In the typical case where $\gamma_1'(0)$ and $\gamma_2'(0)$ are linearly independent vectors, we may restrict to two parameters $\mu \in \mathbb{R}^2$ and transform parameters to $\gamma_1(\mu) = -\mu_1$, $\gamma_2(\mu) = \mu_2$. Furthermore, to allow for stable Turing patterns in the subcritical region $\mu_1 < 0$, we assume $\gamma_4(0) > 0$. We then rescale to $\gamma_4(0) = 1$. The relevant scaling, where we find nontrivial equilibria and periodic orbits, is

$$
A = |\mu_1|^{-1/4}\tilde{A}\mathrm{e}^{\mathrm{i}k_*x}, \; x = |\mu_1|^{1/2}\tilde{x}, \; \mu_2 = |\mu_1|^{1/2}\nu
$$

and $\mu_1 < 0$. Note that we dropped tildes for convenience, although we do not restore the first coordinates introduced on the center manifold! The scaling gives

$$
\begin{aligned}
A_x &= B \\
B_x &= A + \nu A|\tilde{A}|^2 + \gamma_3\mathrm{i}A(\overline{A}B - A\overline{B}) + A|A|^4 + \mathcal{R}(A,B;\mu_1,\nu,x)
\end{aligned}
\tag{27}
$$

and the remainder can be split into

$$
\mathcal{R}(A,B;\mu_1,\nu,x) = |\mu_1|^{1/2}\mathcal{R}_1(A,B;\mu_1,\nu) + |\mu_1|^{m'}\mathcal{R}_2(A,B;\mu_1,\nu,x),
$$

for any fixed $m' > 0$. Here, $\mathcal{R}_1$ corresponds to higher order terms in normal form, whereas $\mathcal{R}_2$ corresponds to the remainder of arbitrary high order, which is not in normal form. The x-dependence of $\mathcal{R}_2$ is rapidly oscillating with period $2\pi|\mu_1|^{1/2}/k_*$.

3.5.3. *Coexistence at the Maxwell point.* We consider the truncated equation

$$\begin{aligned} A_x &= B \\ B_x &= A + \nu A|A|^2 + \gamma_3 \mathrm{i} A(\overline{A}B - A\overline{B}) + A|A|^4. \end{aligned} \tag{28}$$

Note that the subspace $(A, B) \in \mathbb{R}^2 \subset \mathbb{C}^2$ is flow-invariant for the flow to equation (28). The equation in this subspace is Hamiltonian with energy $H(A, B)$ given by

$$H(A,B) = \frac{1}{2}|B|^2 - V(A), \quad V(A) = \frac{1}{2}|A|^2 + \frac{\nu}{4}|A|^4 + \frac{1}{6}|A|^6.$$

Equilibria are given by $A = 0$ and $A_\pm^2 = \frac{1}{2}(-\nu \pm \sqrt{\nu^2 - 4})$. For $\nu > -2$, there is only one equilibrium. At $\nu = -2$, a nontrivial equilibrium $A = 1$ appears. For $\nu < -2$, it splits into two positive equilibria $A_+ > A_- > 0$ (the negative equilibria $-A_\pm$ are conjugate by complex rotation). At the Maxwell point, for $\nu = \nu_{\mathrm{Maxw}} = -4/\sqrt{3}$, A_+ and 0 have equal energy and there exists a unique real trajectory $q(x) = (Q(x), Q'(x)) \in \mathbb{R}^2$ with $Q(x) \to 0$ for $x \to -\infty$ and $Q(x) \to A_+$ for $x \to +\infty$; see Figure 6 for the phase portrait in the real subspace.

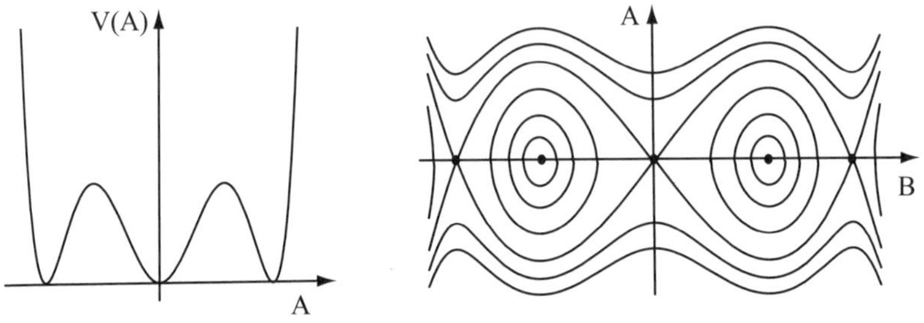

FIGURE 6. *Potential $V(A)$ and phase portrait at the Maxwell point, where Turing patterns and trivial state have equal potential $V(A)$.*

3.5.4. *Transversality.* The goal is to show persistence of the heteroclinic under perturbations of type $\mathcal{R}_1$ in (27). We use transversality as invoked in Proposition 2.24. The equilibria $p_- = (0,0)$ and $p_+ = (A_+, 0)$ continue to equilibria $p_-(\nu) \equiv p_-$ and $p_+(\nu)$ for ν close to ν_{Maxw}. For the (real) four-dimensional dynamics, p_- is hyperbolic with two-dimensional unstable manifold $\mathcal{W}_-^{\mathrm{u}}$. The other equilibrium, p_+ possesses a three-dimensional center-stable manifold $\mathcal{W}_+^{\mathrm{cs}}$. In fact, the linearization at p_+ is hyperbolic within the real subspace (which is flow-invariant) and neutral in the remaining two directions.

PROPOSITION 2.27. *The heteroclinic $q(x)$ for $\nu = \nu_{\mathrm{Maxw}}$ is transverse with respect to the parameter ν:*

$$\mathcal{T}_{\tilde{q}(0)}\left(\bigcup_\nu \mathcal{W}_-^{\mathrm{u}}\right) + \mathcal{T}_{\tilde{q}(0)}\left(\bigcup_\nu \mathcal{W}_+^{\mathrm{cs}}\right) = \mathbb{C}^2 \times \mathbb{R} \tag{29}$$

in the extended phase space $((A, B), \nu) \in \mathbb{C}^2 \times \mathbb{R}$.

Proof. Note that the tangent space to the center manifold $\mathcal{T}_{p_+}\mathcal{W}^{\mathrm{c}}$ spans the complement to the real subspace. Within the flow-invariant real subspace we have transverse intersection of $\bigcup_\nu \mathcal{W}_-^{\mathrm{u}} \cap \mathbb{R}^2$ and $\bigcup_\nu \mathcal{W}_+^{\mathrm{s}}$; the proof is the same as in

Proposition 2.24 for the interface in the case of a cusp. Therefore, the sums of the tangent spaces in (29) is the sum of the real subspace and the direction of the parameter ν from the transverse intersection within the real subspace, and the purely imaginary subspace, from the tangent space to the center manifold $\mathcal{W}^{\mathrm{c}}$ at p_+, which proves the proposition. ∎

COROLLARY 2.28. *Consider equation (27), with* $\mathcal{R}_2 \equiv 0$, *that is, neglecting non normal form terms. Then there is a unique value of* $\nu = \nu_{\mathrm{Maxw}}(\mu_1) = -4/\sqrt{3} + \mathrm{O}(|\mu_1|^{1/2})$, *and a unique periodic orbit* $p_+^k(x) = (P_+^k, (P_+^k)')(x)$ *with* $P_+^k(x) = P_+^0(k)\mathrm{e}^{ikx}$ *for a certain* $P_+^0(k) = |A_+| + \mathrm{O}(k)$, *with* $k = k(\mu_1) = \mathrm{O}(|\mu_1|^{1/2})$, *such that there exists a heteroclinic orbit* $q(x)$ *with* $q(x) \to 0$ *for* $x \to -\infty$, *and* $|q(x) - p_+^k(x)| \to 0$ *for* $x \to +\infty$.

Proof. The stable and unstable manifolds $\bigcup_\nu \mathcal{W}_+^{\mathrm{cs}}$ and $\bigcup_\nu \mathcal{W}_-^{\mathrm{u}}$ are Lipschitz continuously depending on the small perturbation $\mathcal{R}_1$ as smooth manifolds [**Shu87, KH95**]. The transverse intersection therefore persists for an open interval of parameter values μ_1. ∎

3.5.5. *Nonadiabatic effects.* We close the discussion of the Turing instability showing persistence of the interface under the perturbation $\mathcal{R}_2$ in (27), representing non normal form terms, often referred to as nonadiabatic effects (see the discussion below).

PROPOSITION 2.29. *For each* μ_1 *small and all* $\varphi \in \mathbb{R}$, *there exist unique values* $(k, \nu)(\varphi; \mu_1)$, *such that for equation (27) including* $\mathcal{R}_2$, *there is a unique heteroclinic* $q(x)$ *for the parameter value* ν *with* $q(x) \to 0$ *for* $x \to -\infty$ *and* $|q(x) - p_+^k(x+\varphi)| \to 0$ *for* $x \to +\infty$; *see Corollary 2.28 for the definition of* p_+^k.

Proof. We consider $\mathcal{R}_2$ as a perturbation of the normal form system. Note that dependence of $\mathcal{R}_2$ on x is singular. However, following [**FS90**], unstable and center-stable manifolds still depend smoothly on the perturbation. The proof given there for stable and unstable manifolds goes through also for center-stable manifolds as considered here, considering suitable exponentially weighted spaces. Therefore, transverse intersections persist. For the truncated equation, we had a transverse intersection for fixed phase φ: $\varphi = 0$ corresponds to the real equilibrium in the truncated equation. The transverse intersection selects unique values k and ν for the intersection. ∎

Summarizing, we have found stationary patterns of the reaction diffusion system, which consist of stable Turing patterns in the region $x \to +\infty$ and of a stable homogeneous state for $x \to -\infty$. We also showed that, allowing for small variations of the parameter ν, there exists a two-parameter family of such interfaces, parameterized over translation in x and the phase φ of the Turing pattern relative to the interface.

The main difference between the normal form and the full system is that intersections do not necessarily occur along two-dimensional manifolds. The system becomes nonautonomous with $\mathcal{R}_2$ and heteroclinic orbits for *fixed* ν are typically isolated. Indeed, in the full reaction-diffusion system (9), we have an intersection between an unstable manifold of dimension N and a center-stable manifold of dimension $N+1$, with one-dimensional intersection along the heteroclinic. Invoking transversality [**AR**], we find that generically this intersection is transverse and heteroclinics are isolated in phase space, but occur for open ranges in parameter space.

For small amplitudes, this range of parameter values is invisible in the normal form and we expect it to be exponentially small: $\delta\mu_2 \leq \mathrm{Const}_1\, \mathrm{e}^{-\mathrm{Const}_2/\sqrt{|\mu_1|}}$; see for example [**Gel97, Lom00**] for results in this direction.

The phenomenon of standing interfaces between different spatial structures of the PDE to occur for open ranges of parameter values is referred to as a locking phenomenon. It does not occur in 'generic' reaction-diffusion systems between homogeneous equilibria. There, standing interfaces are of codimension one, since stable and unstable manifolds of stable equilibria are of equal dimension N.

A different interpretation of the phenomenon is as follows (see [**Pom86**] for an early reference). The two spatial scales representing wavelengths of Turing patterns and modulations along the heteroclinic are independent in normal form; the temporal dynamics of the PDE quickly relaxes to dynamics involving only long-wave modulations, represented for example by the heteroclinic; nonadiabatic effects illustrate that this relaxation is not perfect and even in long-time dynamics, small spatial scales are still relevant.

Leaving the range of locking, where standing interfaces exist, we expect travelling waves to describe the competition between the trivial state and the stationary periodic pattern. Since in a comoving frame, the periodic patterns are time-periodic, travelling waves cannot be found within the approach taken here. Travelling waves of *stable* Turing-like structures, invading an *unstable* trivial state have been found in [**HS99**]. The analysis there is concerned with the Couette-Taylor problem and shows how stable Taylor vortices invade the unstable Couette flow.

4. Oscillatory bifurcations and spatial dynamics

We consider oscillatory instabilities and look for time-periodic solutions. We therefore rephrase the reaction-diffusion system as a dynamical system in spatial time x, in a phase space of time-periodic functions with fixed period. We then reinterpret the linear instabilities in terms of the linearization of this dynamical system, Section 4.1. We then state a center manifold reduction theorem in Section 4.2 for this infinite-dimensional dynamical system. We conclude with an adaption of the normal form algorithm and a discussion of the reduced dynamics in Section 4.3. The patterns we find are spatially periodic wave trains and interfaces between wave trains and spatially homogeneous equilibria.

4.1. Spatial dynamics on time-periodic functions. Considering spatial dynamics as in (9), the trivial solution $u \equiv 0$ is hyperbolic and isolated at onset, $\mu = 0$, in case of an oscillatory instability (H) or (TH). The center manifold would be trivial. Interesting solutions are time-periodic with temporal period close to the period $2\pi/\omega_*$ predicted by the linear part. Here, ω_* is the imaginary part of the critical eigenvalue.

We study the first-order differential equation

$$\begin{aligned} u_x &= w \\ w_x &= D^{-1}\left(\omega\partial_t u - F(u;\mu)\right) \end{aligned} \tag{30}$$

as an abstract differential equation on the phase space of 2π-time-periodic functions $(u, w)(x, \cdot)$. Note that solutions (u, w) to this differential equation which are bounded on $x \in \mathbb{R}$, uniformly in t, give us $2\pi/\omega$-periodic solutions $U(t, x) = u(x, \omega t)$

of the original reaction-diffusion system (2). We endow the phase space with the topology $(u,w) \in Y = H^{1/2}(S^1,\mathbb{R}^N) \times L^2(S^1,\mathbb{R}^N)$. Functions in Y can be represented by their Fourier series $(u,w)(t) = \sum_{\ell\in\mathbb{Z}}(u^\ell,w^\ell)\mathrm{e}^{\mathrm{i}\ell t}$. The norm on Y can then be defined as

$$|(u,w)|_Y^2 = \sum_{\ell\in\mathbb{Z}} \left((|\ell|+1)|u^\ell|^2 + |w^\ell|^2\right). \tag{31}$$

Since we are interested in nonzero critical eigenvalues, Hypothesis 2.2, we may assume $F(0;\mu) = 0$ without loss of generality. The equation (30) then possesses a trivial solution $(u,w)(x,\cdot) \equiv 0$ for μ close to zero . We may (formally) linearize (30) around this equilibrium of the x-'dynamics' to obtain

$$\begin{aligned} u_x &= w \\ w_x &= D^{-1}\left(\omega\partial_t u - \partial_U F(0;\mu)u\right) \end{aligned} \tag{32}$$

which we briefly write as $\underline{u}_x = \mathcal{A}(\omega;\mu)\underline{u}$. System (32) decouples into a family of equations for the Fourier coefficients (u^ℓ,w^ℓ),

$$\begin{aligned} u_x^\ell &= w^\ell \\ w_x^\ell &= D^{-1}\left(\omega\mathrm{i}\ell u^\ell - \partial_U F(0;\mu)u^\ell\right). \end{aligned} \tag{33}$$

Using this Fourier representation it is straightforward to verify that $\mathcal{A}(\omega;\mu)$ is a closed operator on Y and satisfies the resolvent estimate

$$|(\mathcal{A}(\omega;\mu) - \mathrm{i}k)^{-1}| \le \frac{C}{|k|},$$

with some constant $C > 0$ for all $k \in \mathbb{R}$ with $|k| \ge k_0$. The domain of definition of $\mathcal{A}(\omega;\mu)$ is $Y^1 = H^1(S^1,\mathbb{R}^N) \times H^{1/2}(S^1,\mathbb{R}^N)$, which is compactly embedded into Y. Therefore, the spectrum of $\mathcal{A}(\omega;\mu)$ consists of isolated eigenvalues of finite multiplicity. However, $\mathrm{Re}(\mathrm{spec}\,\mathcal{A}(\omega;\mu))$ is unbounded in $\mathbb{R}^+$ and $\mathbb{R}^-$, and the initial value problem to (32) is ill-posed.

With the Fourier decomposition (33), we can (explicitly) split $Y = E^{\mathrm{s}} \oplus E^{\mathrm{c}} \oplus E^{\mathrm{u}}$ such that the subspaces E^j, $j = \mathrm{s}$, c, and u are invariant under $\mathcal{A}_0 := \mathcal{A}(\omega_*;0)$ and the spectrum decomposes into $\mathrm{spec}\,\mathcal{A}_0|_{E^{\mathrm{s}}} \subset \mathbb{C}^-$, $\mathrm{spec}\,\mathcal{A}_0|_{E^{\mathrm{u}}} \subset \mathbb{C}^+$, and $\mathrm{spec}\,\mathcal{A}_0|_{E^{\mathrm{c}}} \subset \mathrm{i}\mathbb{R}$. Denote again by P^{s}, P^{c}, and P^{u} the spectral projections according to this decomposition. Note that $\dim E^{\mathrm{c}} < \infty$, but $\dim E^{\mathrm{s}} = \dim E^{\mathrm{u}} = \infty$.

The abstract differential equation possesses several important symmetries. Reflectional symmetry $x \to -x$ of the original reaction-diffusion system induces reversibility. The equation is invariant under the action of R and spatial time-reversal $x \to -x$, where the involution R acts according to $R: Y \to Y$, $(u,w) \mapsto (u,-w)$. In particular, the linearization anti-commutes, $\mathcal{A}_0 R = -R\mathcal{A}_0$ on Y^1, and eigenspaces are permuted, $RE^{\mathrm{c}} = E^{\mathrm{c}}$, $RE^{\mathrm{u}} = E^{\mathrm{s}}$. In addition, there is the one-parameter group $\mathcal{SO}(2)$ of rotations acting on Y via $\gamma_\theta : (u,w)(t) \mapsto (u,w)(t-\theta)$, with $\theta \in \mathbb{R}/2\pi\mathbb{Z} \sim \mathcal{SO}(2)$. Since the original reaction-diffusion system was autonomous and since we linearize along a temporal equilibrium state, the linearization commutes with temporal time shift, $\gamma_\theta\mathcal{A}(\omega;\mu) = \mathcal{A}(\omega;\mu)\gamma_\theta$, and eigenspaces are invariant, $\gamma_\theta E^j = E^j$ for $j = \mathrm{s}$, c, and u.

Temporal time-shift γ and spatial reflection commute, such that $\gamma_\theta R = R\gamma_\theta$, for all $\theta \in \mathbb{R}/2\pi\mathbb{Z}$.

LEMMA 2.30. *Assume Hypotheses 2.2, 2.5, and 2.6 for an oscillatory instability (H). We then have* $\dim E^{\mathrm{c}} = 4$ *with eigenvalue* $\nu = 0$ *geometrically double. In suitable, complex, coordinates* $(A, B, \overline{A}, \overline{B})$*, we have the following representations for* $\mathcal{A}_0$*,* R*, and* γ_θ*:*

$$\mathcal{A}_0|_{E^c} = \begin{pmatrix} 0 & 1 & 0 & 0 \\ 0 & 0 & 0 & 0 \\ 0 & 0 & 0 & 1 \\ 0 & 0 & 0 & 0 \end{pmatrix},$$

and

$$\begin{aligned} R(A, B, \overline{A}, \overline{B}) &= (A, -B, \overline{A}, -\overline{B}) \\ \gamma_\theta(A, B, \overline{A}, \overline{B}) &= (\mathrm{e}^{\mathrm{i}\theta} A, \mathrm{e}^{\mathrm{i}\theta} B, \mathrm{e}^{-\mathrm{i}\theta}\overline{A}, \mathrm{e}^{-\mathrm{i}\theta}\overline{B}). \end{aligned}$$

A solution $U(t,x)$ *of the original linearized reaction-diffusion system (3) is then reconstructed from* $U(t,x) = A(x)e^{\mathrm{i}\omega_* t}U_0 + \text{c.c.}$ *with* U_0 *the Hopf eigenvector of the kinetics* $(\partial_U F(0;0) - \mathrm{i}\omega_*)U_0 = 0$.

The proof is similar to the proof of Lemma 2.11. Bounded solutions to (33) only exist when $\ell = 1$ or $\ell = -1$. We then have to discuss bounded solutions to this linear ordinary differential equation like in Lemma 2.11. We omit the details, which are completely analogous to the proof of Lemma 2.11.

We conclude with a brief discussion of the Turing-Hopf instability, although we do not pursue this case any further in later chapters. The case of a Turing-Hopf instability (TH) is analyzed in [**IM91**], in the context of instabilities in fluid flows. The dimension of the critical center eigenspace is $\dim E^c = 4$ with algebraically simple eigenvalue $\nu = \mathrm{i}k_*$. Principal vectors are excluded since we fixed the temporal period $2\pi/\omega_*$, and the frequency of the eigenfunction $\omega = \operatorname{Im}\lambda$ depends on the wave number k, with nonzero derivative c_g, see Hypothesis 2.7. In suitable, complex, coordinates $A_1, A_2, \overline{A_1}, \overline{A_2}$, we have the following representation for $\mathcal{A}_0$, R, and γ_θ

$$\mathcal{A}_0|_{E^c} = \begin{pmatrix} \mathrm{i}k_* & 0 & 0 & 0 \\ 0 & -\mathrm{i}k_* & 0 & 0 \\ 0 & 0 & \mathrm{i}k_* & 0 \\ 0 & 0 & 0 & -\mathrm{i}k_* \end{pmatrix},$$

and

$$R(A_1, A_2, \overline{A_1}, \overline{A_2}) = (A_2, A_1, \overline{A_2}, \overline{A_1}),$$
$$\gamma_\theta(A_1, A_2, \overline{A_1}, \overline{A_2}) = (\mathrm{e}^{\mathrm{i}\theta} A_1, \mathrm{e}^{\mathrm{i}\theta} A_2, \mathrm{e}^{-\mathrm{i}\theta}\overline{A_1}, \mathrm{e}^{-\mathrm{i}\theta}\overline{A_2}).$$

A solution $U(t,x)$ of the original linearized reaction-diffusion system (3) is then reconstructed from $A_1(x)e^{\mathrm{i}(\omega_* t + k_* x)}U_0 + A_2(x)e^{\mathrm{i}(\omega_* t - k_* x)}U_0 + \text{c.c.}$ Here, U_0 denotes the Hopf eigenvector, $(-Dk_*^2 + \partial_U F(0;0) - \mathrm{i}\omega_*)U_0 = 0$.

A nonsemisimple linear part — leading to a more complicated normal form and various types of nonlinear defects — is recovered in a coordinate frame moving with speed $\pm c_\mathrm{g}$. However, typically the second eigenfunction A_2 will then have a frequency different from the frequency of the first mode and the center eigenspace will again be complex two-dimensional. Only in the degenerate case of vanishing group velocity, where Hypothesis 2.7 is violated, we can find an 8-dimensional center eigenspace; see [**RK00**] for an analysis in this direction.

We also mention [**HSS99**], where defect-like solutions are studied for coupled complex Ginzburg-Landau equations, which are shown to model counter-propagating waves, as created in the Turing-Hopf instability.

4.2. Center manifolds. Bounded solutions of the linearized problem are described by a finite-dimensional ODE on E^{c}. The same is true for small bounded solutions of the nonlinear problem (30), which we write in the abstract form

$$\underline{u}_x = \mathcal{A}(\omega;\mu)\underline{u} + \mathcal{F}(\underline{u};\mu), \tag{34}$$

with

$$\begin{aligned} \mathcal{A}(\omega;\mu)(u,w)^T &= \left(w, D^{-1}(\omega\partial_t u - \partial_U F(0;\mu)u)\right)^T \\ \mathcal{F}((u,w);\mu) &= (0, -D^{-1}(F(u;\mu) - \partial_U F(0;\mu)u))^T. \end{aligned}$$

The operator $\mathcal{A}$ is closed on Y with domain of definition Y^1. The nonlinearity $\mathcal{F}$ maps Y^1 into Y^1 smoothly, since $F : H^1(S^1) \to H^{1/2}(S^1)$, $u(\cdot) \mapsto F(u(\cdot))$ is smooth from the embedding $H^1(S^1) \to C^0(S^1)$.

We say $\underline{u}(\cdot) \in C^0(J,Y)$ is a solution on the interval J, if $\underline{u}(\cdot) \in C^0(\operatorname{int} J, Y^1) \cap C^1(\operatorname{int} J, Y)$ and $\underline{u}(x)$ satisfies (34) for $x \in \operatorname{int} J$. Recall the Definition 2.13 of a center manifold.

THEOREM 2.31. *Assume that the reaction-diffusion system undergoes an oscillatory instability: a homogeneous Hopf (H), or a Turing-Hopf (TH) instability. Let E^{c} denote the center eigenspace of $\mathcal{A}_0$.*

Then, for any $0 < m < \infty$, there is a $\delta > 0$, such that there exists a C^m-center manifold $\mathcal{W}^{\mathrm{c}}$ with reduced flow $\Phi^{\mathrm{c}}(x,\underline{u})$, for $|\mu| + |\omega - \omega_*| < \delta$. For $\mu = 0$ and $\omega = \omega_*$, $\mathcal{W}^{\mathrm{c}}$ is tangent to E^{c} in $\underline{u} = 0$. The center manifold $\mathcal{W}^{\mathrm{c}}$ and the reduced flow $\Phi^{\mathrm{c}}(\cdot,\cdot)$ depend C^m on μ and ω. Center manifold and reduced flow preserve the symmetry of the original equation. In particular, $R\mathcal{W}^{\mathrm{c}} = \mathcal{W}^{\mathrm{c}}$, and $R\Phi^{\mathrm{c}}(x,\underline{u}) = \Phi^{\mathrm{c}}(-x, R\underline{u})$, whenever one of both is defined. Also, $\gamma_\theta\mathcal{W}^{\mathrm{c}} = \mathcal{W}^{\mathrm{c}}$ and $\gamma_\theta\Phi^{\mathrm{c}}(x,\underline{u}) = \Phi^{\mathrm{c}}(x,\gamma_\theta\underline{u})$, for all $\theta \in \mathbb{R}/2\pi\mathbb{Z}$.

The proof goes back, in spirit, to [**Kir82**]. The time-periodic case was first treated in [**IM91**]. For a general reference, see [**VI91**].

Smooth dependence on ω, which appears in the leading order part of the equation, follows from results on center manifolds for quasi-linear equations [**Mie88a**]. In the particular case presented here, a simple rescaling provides an alternative proof. Set $x = \sqrt{\frac{\omega_*}{\omega}}\tilde{x}$, $w = \sqrt{\frac{\omega}{\omega_*}}\tilde{w}$, and $F(u;\mu) = \frac{\omega}{\omega_*}\tilde{F}(u;\mu,\omega)$. Then (30) becomes

$$\begin{aligned} u_{\tilde{x}} &= \tilde{w} \\ \tilde{w}_{\tilde{x}} &= D^{-1}\left(\omega_*\partial_t u - \tilde{F}(u;\mu,\omega)\right) \end{aligned}$$

with center manifold $\tilde{\mathcal{W}}^{\mathrm{c}}$ and local flow $\tilde{\Phi}^{\mathrm{c}}$, depending smoothly on ω. Going back through the scalings shows that center manifold $\mathcal{W}^{\mathrm{c}}$ and local flow Φ^{c} depend smoothly on ω.

4.3. Normal forms, reduced equations, and coexistence. There is an analogue of Proposition 2.16 for the unfolding of the linear part, here in the oscillatory case.

PROPOSITION 2.32. *Consider a linearly generically unfolded Hopf (H) instability. In smooth coordinates, (μ,ω)-close to the coordinates of Lemma 2.30 on E^{c}, we obtain the reduced equation for the complex amplitudes A, B*

$$\begin{aligned} A_x &= B + \mathrm{O}\left((|A| + |B|)^2\right) \\ B_x &= \gamma_1(\mu,\omega)A + \mathrm{O}\left((|A| + |B|)^2\right) \end{aligned} \tag{35}$$

with $\gamma_1(\mu,\omega)\in\mathbb{C}$, $\gamma_1(0,\omega_*)=0$, $\mathrm{Re}\,\partial_\mu\gamma_1(0,\omega_*)\neq 0$, *and* $\mathrm{Im}\,\partial_\omega\gamma_1(0,\omega_*)\neq 0$.

Proof. The linear part is automatically in the described normal form, due to equivariance and reversibility, up to a rescaling of B. The form of γ_1 is a consequence of the genericity of the unfolding, Definition 2.10. ∎

The next step is a nonlinear normal form. The nonsemisimple part in the Hopf case is the same as for the Turing instability. The action of the continuous normal form symmetry group generated by the semi-simple part in the Turing instability is the same as the action of the time-shift symmetry γ_θ. The only difference lies in the action of reversibility, without complex conjugation. From [**ETBCI87**] we easily obtain, restricting to reversible polynomials, the following normal form:

$$\begin{aligned} A_x &= B+\mathrm{i}A(A\overline{B}-B\overline{A})P_1\left(|A|^2,(A\overline{B}-B\overline{A})^2;\mu,\omega\right) \\ B_x &= \gamma_1(\mu,\omega)A+AP_2\left(|A|^2,(A\overline{B}-B\overline{A})^2;\mu,\omega\right)+ \\ &\quad +\mathrm{i}B(A\overline{B}-B\overline{A})P_1\left(|A|^2,(A\overline{B}-B\overline{A})^2;\mu,\omega\right) \end{aligned} \tag{36}$$

with P_1 and P_2 being complex polynomials in their arguments, which vanish in the origin $P_j(0,0;\mu,\omega)=0$, $j=1,2$.

We write $\gamma_1=-\mu_1+\mathrm{i}\hat{\omega}$ and $\partial_1P_2(0,0,0,\omega_*)=\gamma_2$. We then scale $A=|\mu_1|^{1/2}\tilde{A}$, $x=|\mu_1|^{-1/2}\tilde{x}$, $B=|\mu_1|\tilde{B}$, and $\hat{\omega}=|\mu_1|\tilde{\omega}$ and arrive at

$$\begin{aligned} \tilde{A}_{\tilde{x}} &= \tilde{B}+\mathrm{O}\left(|\mu_1|^{1/2}\right) \\ \tilde{B}_{\tilde{x}} &= (\pm 1+\mathrm{i}\tilde{\omega})\tilde{A}+\gamma_2\tilde{A}|\tilde{A}|^2+\mathrm{O}\left(|\mu_1|^{1/2}\right) \end{aligned} \tag{37}$$

If $\gamma_2\neq 0$, the truncated equation is the steady-state equation for the cubic complex Ginzburg-Landau equation

$$\tilde{A}_{\tilde{t}}=(1+\beta_1)A_{\tilde{x}\tilde{x}}+(\pm 1+\mathrm{i}\beta_2)\tilde{A}+(\pm 1+\mathrm{i}\beta_3)\tilde{A}|\tilde{A}|^2, \tag{38}$$

with real coefficients β_j, after rescaling $\tilde{A}$.

Besides periodic solutions of the form $\tilde{A}(\tilde{x})=a\mathrm{e}^{\mathrm{i}k\tilde{x}}$, there may exist many localized solutions, asymptotic to zero or to periodic solutions. Countably many localized solutions can be constructed close to the degenerate case of real coefficients $\beta_1=\beta_2=\beta_3=0$; see [**Doe96**]. Some of the localized solutions can be computed explicitly, see [**vSH92**].

If the complex cubic coefficient vanishes, $\gamma_2=0$ at $\mu=0$, we assume that this degeneracy is unfolded by the parameters μ_2 and μ_3 and set $\partial_1P_2(0,0;\mu,\omega_*)=\mu_2+\mathrm{i}\mu_3$, possibly transforming parameter space μ,ω by a local diffeomorphism. We then scale $\mu_2=\tilde{\mu}_2|\mu_1|^{1/2}$, $\mu_3=\tilde{\mu}_3|\mu_1|^{1/2}$, $A=|\mu_1|^{1/4}\tilde{A}$, $x=|\mu_1|^{-1/2}\tilde{x}$, $B=|\mu_1|^{3/4}\tilde{B}$, and $\hat{\omega}=|\mu_1|\tilde{\omega}$ and arrive at

$$\begin{aligned} \tilde{A}_{\tilde{x}} &= \tilde{B}+\mathrm{O}\left(|\mu_1|^{1/2}\right) \\ \tilde{B}_{\tilde{x}} &= (\pm 1+\mathrm{i}\tilde{\omega})\tilde{A}+(\tilde{\mu}_2+\mathrm{i}\tilde{\mu}_3)\tilde{A}|\tilde{A}|^2+\gamma_3\tilde{A}|\tilde{A}|^4+\mathrm{O}\left(|\mu_1|^{1/2}\right) \end{aligned} \tag{39}$$

with $\gamma_3=\partial_1^2P_2$ in zero. This is the cubic-quintic Ginzburg-Landau equation, which for certain parameter values may possess stable standing pulses and fronts, see [**vSH92, KS98**]. In particular, in the weakly subcritical regime, there may exist standing fronts between the stable trivial state and patterns of the form $\tilde{A}(\tilde{x})=a\mathrm{e}^{\mathrm{i}k\tilde{x}}$.

We illustrate coexistence in oscillatory instabilities starting from the real cubic-quintic Ginzburg-Landau equation, Section 3.5, equation (28) with $\gamma_2 = 0$:

$$A_{xx} = (1 + \mathrm{i}\omega)A + (\nu + \mathrm{i}\gamma_1)A|A|^2 + (1 + \mathrm{i}\gamma_2)A|A|^4, \tag{40}$$

where $\gamma_1 = \gamma_2 = \omega = 0$. For $\gamma_1 = \gamma_2 = 0$, we showed in Section 3.5 that, at the Maxwell point $\nu = -4/\sqrt{3}$, there exists a heteroclinic orbit connecting $A = 0$ to $A = A_+$, with $A_+^2 = (-\nu + \sqrt{\nu^2 - 4})/2 = 2\sqrt{3}$. The heteroclinic is transverse as a heteroclinic to $\{\mathrm{e}^{\mathrm{i}\varphi}A_+;\ \varphi \in \mathbb{R}\}$ in the extended phase space with the parameter ν added. Varying ω, the transverse intersection persists and we find transverse heteroclinic orbits to periodic orbits $A\mathrm{e}^{\mathrm{i}k(\omega)x}$ near $A = A_+$, $k = 0$. Since the perturbation $\mathrm{i}\omega A$ gives the only contribution to the vector field in the imaginary subspace, with definite sign, $\mathrm{sgn}\,(\omega A) = \mathrm{sgn}\,(\omega)$, the selected wavenumber changes nontrivially with ω, $\frac{\mathrm{d}k}{\mathrm{d}\omega} \neq 0$. In other words, given k close to zero, we find $(\omega_*, \nu_*)(k)$ and a unique heteroclinic connecting 0 to $a(k)\mathrm{e}^{\mathrm{i}kx}$ for values of the parameters $(\omega, \nu) = (\omega_*, \nu_*)(k)$ and $\gamma_1 = \gamma_2 = 0$. Adding the perturbation γ_1, $\gamma_2 \neq 0$, the family of periodic orbits $a\mathrm{e}^{\mathrm{i}kx}$ persists as a normally hyperbolic invariant manifold [**Fen79**]. Transverse heteroclinics now correspond to transverse intersections between the unstable manifold of the origin and any fixed strong stable fiber of the normally hyperbolic invariant manifold, close to the real subspace, adjusting the parameters ν and ω appropriately. Periodic orbits of the form $a\mathrm{e}^{\mathrm{i}kx}$ satisfy the algebraic relations

$$\omega = -\gamma_1 a^2 - \gamma_2 a^4, \quad k^2 = -1 - \nu a^2 - a^4. \tag{41}$$

For $a \leq A_+ = 2\sqrt{3}$, k^2 is positive and $\omega = -\gamma_1 a^2 - \gamma_2 a^4$ and $\frac{\mathrm{d}\omega}{\mathrm{d}a} = 2\gamma_1 2\sqrt{3} - 4\gamma_2(2\sqrt{3})^3$ typically carry definite signs. Since $\frac{\mathrm{d}k^2}{\mathrm{d}a}|_{a=2\sqrt{3}} < 0$, the group velocity $\frac{\mathrm{d}\omega}{\mathrm{d}k} = \frac{\mathrm{d}\omega}{\mathrm{d}a} \cdot \frac{\mathrm{d}a}{\mathrm{d}k}$ can take both positive and negative signs, depending precisely upon the sign of k. As a consequence of this discussion, we conclude that the change of the orbit $a\mathrm{e}^{\mathrm{i}kx}$ with ω is small. We may therefore adjust ω and ν to find a connecting orbit from zero to any periodic orbit with $k \sim 0$ small and group velocity $\frac{\mathrm{d}\omega}{\mathrm{d}k} > 0$:

PROPOSITION 2.33. *Consider a linearly generically unfolded Hopf bifurcation, with degenerate cubic coefficients in the truncated quintic normal form; see (40), with γ_1, γ_2 close to zero, and ν close to $-4/\sqrt{3}$. Then for all waves $A_+\mathrm{e}^{\mathrm{i}kx}$ with positive group velocity $c_\mathrm{g} = \frac{\mathrm{d}\omega}{\mathrm{d}k}$ determined from (41), there exists ν close to $-4/\sqrt{3}$ such that there exists a heteroclinic solution to (40), asymptotic to zero for $x \to -\infty$ and to $A_+\mathrm{e}^{\mathrm{i}kx}$ for $x \to +\infty$. The solutions are transverse in the extended phase space and persist as a branch of solutions for the full reaction-diffusion system (30).*

Note that the persistence under higher order perturbations which are not in normal form is less subtle than in the case of a Turing instability since the continuous $\mathcal{SO}(2)$-symmetry is induced by temporal time-shift and is present in the full system, not only in the normal form.

Note however, that we have to adjust the parameter ν in the equation to find a standing interface — just like in the case of an interface between spatially homogeneous states in the cusp bifurcation.

The case of negative group velocity is simpler. It is not difficult to verify that periodic orbits with negative group velocity are stable inside the Fenichel slow manifold. Heteroclinic orbits therefore occur for intervals of parameter values ν.

The above case is a codimension three situation since both real, and imaginary part of the cubic coefficient vanish simultaneously. The transition between sub- and supercritical bifurcations, where only the real part of the cubic coefficient crosses zero does not seem to be well understood. We refer to [**EI89**] for a partial discussion of this problem. No results on fronts and pulses seem to exist for this case, which leads to a small perturbation of (37) with γ_2 purely imaginary.

REMARK 2.34. *In a slightly more general spirit, we might have tried to find time-periodic solutions in a coordinate system* $\xi = x - ct$*, moving with speed* c*; see Remark 2.26 for the case of a stationary instability. We recover the wave speed* c *as a parameter in the linear part of the reduced equation. Similarly to the case of a stationary instability, we find a new term* $-\overline{\gamma}cB$ *in the equation for* B_ξ *with* $\mathrm{Re}\,\overline{\gamma} > 0$*. In particular, the heteroclinic orbits representing the coexistence patterns that we described above, derived as perturbations from the real cubic-quintic Ginzburg-Landau equation, are transverse intersections in the parameter* c*. Therefore, in addition to the codimension-one surface of standing coexistence interfaces, we find moving interfaces, where either of the stable states, oscillatory or homogeneous, may invade the other.*

CHAPTER 3

Stationary radially symmetric patterns

We extend the general reduction procedure from Chapter 2 to higher space dimensions, restricting to radially symmetric solutions. First, the problem is formulated as a nonautonomous dynamical system, and the linearized differential equation is analyzed in terms of exponential dichotomies in Section 1. Section 2 contains the main reduction results. We first prove existence of an r-dependent, dynamically invariant center manifold in Section 2.1. We then enlarge this manifold in the far-field, r large, to find an asymptotic center manifold with r-independent tangent space. Within this enlarged manifold, we then proceed in Section 3 to compute Taylor jets of the reduced vector field and derive a normal form algorithm, taking care of curvature terms, represented by dependence of the vector field on powers of $1/r$. The main results are presented in Section 4. First, the far-field equations are scaled in a long-wavelength expansion and universal, reduced equations are derived, Section 4.1. We then find bounded solutions in these universal equations as transverse heteroclinic orbits, Section 4.2. Section 4.3 shows how to construct solutions of the full reaction-diffusion system from the universal, reduced far-field equations. In particular, a rigorous matching-procedure with the center of the pattern gives us existence of branches of solutions, bifurcating from the origin; Theorems 3.18, 3.19, and 3.20.

1. Classification and radial dynamics

We study instabilities in higher space dimensions, focusing on radially symmetric solutions. Consider therefore

$$U_t = D\triangle_x U + F(U;\mu), \tag{42}$$

with $U \in \mathbb{R}^N$, $\mu \in \mathbb{R}^p$, and $x = (x_1, \dots, x_n) \in \mathbb{R}^n$. We assume existence of a trivial, spatially homogeneous equilibrium $F(0;0) = 0$. The linearization becomes

$$V_t = D\triangle_x V + \partial_U F(0;0)V. \tag{43}$$

As in Chapter 2, Section 1, Fourier transformation leads to a classification of possible instabilities: critical wave vectors are of the form $\mathrm{e}^{\mathrm{i}((k_*,x)+\omega_* t)}U_0$. We obtain the four cases (O), (T), (H), and (TH), as in Definition 2.3. Formally, when referring to an instability of type (T), say, in higher space dimension, we restrict equation (42) to one-dimensional solutions, which do not depend on $x_2, \dots, x_n$, and we then apply the definitions from Chapter 2, Section 1.

Note however that due to rotational invariance of equation (43), whenever $\mathrm{e}^{\mathrm{i}((k_*,x)+\omega_* t)}U_0$ is a solution to (43), $\mathrm{e}^{\mathrm{i}((k,x)+\omega_* t)}U_0$ is a solution, too, whenever $|k| = |k_*|$. In the particular case of a Turing bifurcation, the kernel of the linearization is infinite-dimensional at onset of instability, leading to notorious complications even in formal derivations of amplitude equations [**NW69, Schn95**].

In this chapter, we only treat the cases (O) and (T). We consider generically unfolded stationary instabilities, Definition 2.10, and look for small, bounded, radially symmetric solutions. In polar coordinates, we may rewrite the stationary reaction-diffusion equation $D\Delta_x U + F(U;\mu) = 0$ as a nonautonomous, singular, ordinary differential equation in the radius r,

$$\begin{aligned} u_r &= w \\ w_r &= -\frac{n-1}{r}w - D^{-1}F(u;\mu), \end{aligned} \tag{44}$$

on the phase space $Y = \mathbb{R}^{2N}$. Note that the equation is defined for $0 < r < \infty$.

We will show that there are well-defined limiting equations in $r = 0$ and in $r = \infty$. Formally, the equation is well-defined on $r < 0$ as well. Moreover, we may reflect $(u, w, r) \mapsto (u, -w, -r)$, which leaves the equation invariant. In a neighborhood of $r = \infty$, this reversibility operation imposes severe restrictions on the possible dynamics, as we will see in Section 3. In this sense, reversibility is restored.

In a more compact notation, we consider

$$\underline{u}_r = \mathcal{A}(r)\underline{u} + \mathcal{F}(\underline{u};\mu), \tag{45}$$

where

$$\mathcal{A}(r) = \begin{pmatrix} 0 & 1 \\ -D^{-1}\partial_U F(0;0) & -\frac{n-1}{r} \end{pmatrix},$$

and

$$\mathcal{F}(\underline{u};\mu) = \begin{pmatrix} 0 \\ -D^{-1}(F(u;\mu) - \partial_U F(0;0)u) \end{pmatrix}.$$

The linearization about the trivial solution $(u, w)(r) \equiv 0$ is

$$\begin{aligned} v_r &= w \\ w_r &= -\frac{n-1}{r}w - D^{-1}\partial_U F(0;0)v, \end{aligned} \tag{46}$$

or, $\underline{v}_r = \mathcal{A}(r)\underline{v}$. For r close to zero, we consider a slow spatial time variable $\tau = \log r$, which removes the singular behavior in $r = 0$:

$$\begin{aligned} u_\tau &= \mathrm{e}^\tau w \\ w_\tau &= -(n-1)w - \mathrm{e}^\tau D^{-1}F(u;\mu), \end{aligned} \tag{47}$$

with linearization

$$\begin{aligned} v_\tau &= \mathrm{e}^\tau w \\ w_\tau &= -(n-1)w - \mathrm{e}^\tau D^{-1}\partial_U F(0;0)v. \end{aligned} \tag{48}$$

Both equations (46) and (47) can be rewritten as autonomous equations in an extended phase space, adding an equation for spatial time. For $r \to \infty$, we rewrite

$$\begin{aligned} u' &= w \\ w' &= -(n-1)\alpha w - D^{-1}F(u;\mu) \\ \alpha' &= -\alpha^2 \end{aligned} \tag{49}$$

where $'$ denotes differentiation with respect to $r = 1/\alpha$. For $r \to 0$, the following form is more convenient:

$$\begin{aligned} \dot{u} &= rw \\ \dot{w} &= -(n-1)w - rD^{-1}F(u;\mu) \\ \dot{r} &= r, \end{aligned} \tag{50}$$

where $\dot{}$ now denotes differentiation with respect to τ.

The subspace $r = 0$ in (50) is a flow-invariant subspace with linear flow

$$\begin{aligned} \dot{u} &= 0 \\ \dot{w} &= -(n-1)w, \end{aligned} \tag{51}$$

containing an N-dimensional normally hyperbolic subspace of equilibria $w = 0$.

The subspace $\alpha = 0$, corresponding to $r = \infty$, is flow-invariant for (49). Within this subspace, we recover the one-dimensional spatial dynamics from Chapter 2,

$$\begin{aligned} u' &= w \\ w' &= -D^{-1}F(u;\mu), \end{aligned} \tag{52}$$

with linearization

$$\begin{aligned} v' &= w \\ w' &= -\partial_U F(0;0)v, \end{aligned} \tag{53}$$

which we write as $\underline{v}' = \mathcal{A}(\infty)\underline{v}$.

For the asymptotic equations of the linearized problem, we can define a hyperbolic splitting as follows. Denote by $E_-^{\text{cu}} = \{(u,w); w = 0\}$ and $E_-^{\text{s}} = \{(u,w); u = 0\}$ the flow-invariant linear center-unstable and stable subspaces for (51). Similarly let E_+^{s}, E_+^{c}, and E_+^{u} denote center, stable, and unstable subspaces of the linear equation (53). Note that E_+^{c} is the center space we considered in the one-dimensional problem in Chapter 2, Lemma 2.11.

For the nonautonomous equation, these subspaces continue to r-dependent subspaces, which are characterized by asymptotic exponential growth of solutions starting within these subspaces.

LEMMA 3.1. *The linearized equation (48) possesses an exponential dichotomy on $\tau \leq 0$, that is, there are linear evolution operators $\Phi_-^{\text{cu}}(\tau,\sigma)$, $\Phi_-^{\text{s}}(\tau,\sigma)$, on the phase space $Y = \mathbb{R}^{2N}$, and a constant $C > 0$, such that the following holds. The trajectories $\Phi_-^j(\cdot,\sigma)\underline{u}$ are a solution of (48) for all $\underline{u} \in Y$ and σ fixed, and*

(i) $\Phi_-^j(\tau,\sigma)\Phi_-^j(\sigma,\rho) = \Phi_-^j(\tau,\rho)$, *for all $\tau \leq \sigma \leq \rho \leq 0$ if $j = \text{cu}$, and for all $\rho \leq \sigma \leq \tau \leq 0$ if $j = \text{s}$;*
(ii) $|\Phi_-^{\text{cu}}(\tau,\sigma)| \leq C$, *for $\tau \leq \sigma \leq 0$;*
(iii) $|\Phi_-^{\text{s}}(\tau,\sigma)| \leq Ce^{-|\tau-\sigma|}$, *for $\sigma \leq \tau \leq 0$;*
(iv) $\Phi_-^{\text{s}}(\tau,\tau) + \Phi_-^{\text{cu}}(\tau,\tau) = \text{id}$, *for all $\tau \leq 0$.*

Equation (46) possesses an exponential dichotomy on $r \geq 1$. More precisely, there are evolution operators $\Phi_+^{\text{u}}(r,s)$, $\Phi_+^{\text{c}}(r,s)$, $\Phi_+^{\text{s}}(r,s)$, defined on the phase space $Y = \mathbb{R}^{2N}$, and constants $\eta_+^{\text{u}}, \eta_+^{\text{s}} > 0$, and for any $\eta_+^{\text{c}} > 0$, there is a constant $C > 0$ such that the following holds. The trajectories $\Phi_+^j(\cdot,s)\underline{u}$ are a solution of (46) for all $\underline{u} \in Y$ and s fixed, and

(i) $\Phi_+^j(r,s)\Phi_+^j(s,s') = \Phi_+^j(r,s')$, *for all $1 \leq r \leq s \leq s'$, $j = \text{u}$, and $1 \leq s' \leq s \leq r$, $j = \text{s}$, and $0 \leq s', s, r$, $j = \text{c}$;*

(ii) $|\Phi^{\mathrm{u}}_{+}(r,s)| \le C\mathrm{e}^{-\eta^{\mathrm{u}}_{+}|r-s|}$, *for* $1 \le r \le s$;
(iii) $|\Phi^{\mathrm{s}}_{+}(r,s)| \le C\mathrm{e}^{-\eta^{\mathrm{s}}_{+}|r-s|}$, *for* $1 \le s \le r$;
(iv) $|\Phi^{\mathrm{c}}_{+}(r,s)| \le C\mathrm{e}^{\eta^{\mathrm{c}}_{+}|r-s|}$, *for* $1 \le s, r$;
(v) $\Phi^{\mathrm{s}}_{+}(r,r) + \Phi^{\mathrm{c}}_{+}(r,r) + \Phi^{\mathrm{u}}_{+}(r,r) = \mathrm{id}$, *for all* $r \ge 1$.

We refer to [**Cop78, San93, PSS97**] for a discussion of exponential dichotomies, including proofs of the above lemma.

The operators $\Phi^{j}_{\pm}(s,s)$ are projections for any fixed s, since $(\Phi^{j}_{\pm}(s,s))^2 = \Phi^{j}_{\pm}(s,s)$ from (i). We denote by $E^{j}_{\pm}(s) := \mathrm{Rg}\,(\Phi^{j}_{\pm}(s,s))$ the ranges, which are time-dependent linear subspaces.

2. Center manifolds

2.1. The nonautonomous center manifold $\tilde{\mathcal{W}}^{\mathrm{c}}$. In the spirit of Chapter 2, we construct a locally invariant manifold for the dynamical system (49), (50), in the extended phase space. The manifold we construct contains all small solutions, bounded for all spatial times $r \ge 0$. Roughly speaking, it is obtained as the intersection of the set of solutions to (50), bounded for $r \to 0$, and the set of solutions to (49), bounded for $r \to \infty$. Since the equation is nonautonomous, this manifold depends on r.

As we already pointed out, E^{c}_{+} denotes the generalized center eigenspace of the (asymptotic) one-dimensional problem, $\underline{v}' = \mathcal{A}(\infty)\underline{v}$, obtained at $r = +\infty$. In particular, $\mathcal{A}(\infty)$ anti-commutes with the involution $R : (u,w) \mapsto (u,-w)$: the asymptotic equation is reversible. As a consequence, $\dim E^{\mathrm{c}}_{+}$ is even.

Notation: We write $E \oplus F \le Y$ for sums of subspaces within the phase space Y, and we write $E \times J \subset Y \times \mathbb{R}^{+}$ for product subsets of the extended phase space. We denote locally invariant manifolds with the letters $\tilde{\mathcal{W}} \subset Y \times \mathbb{R}^{+}$, with superscripts u, c, s indicating exponential growth or decay of solutions inside the manifold, and with possible subscripts $+,-$ indicating whether the manifold is contained in $r > 1$ or $r < 1$, respectively. We also write $\mathcal{W}(r)$ for time slices $\tilde{\mathcal{W}} \cap (Y \times \{r = r_0\})$, that is, we drop the tilde when we refer to the manifold as a subset of Y only.

We adapt the definition of a local center manifold to the nonautonomous setting used here.

DEFINITION 3.2. [Nonautonomous center manifold] A C^m-center manifold $\tilde{\mathcal{W}}^{\mathrm{c}}$ for the radial dynamics (45), close to the origin $\{(\underline{u}, r); \underline{u} = 0,\ r \in \mathbb{R}^{+}\}$ in the extended phase space $Y \times \mathbb{R}^{+}$ of (50), given as a fiber-bundle over time r, $\bigcup_{r>0} \mathcal{W}^{\mathrm{c}}(r) \times \{r\} \subset Y \times \mathbb{R}^{+}$, is an open C^m-manifold together with a local flow $\tilde{\Phi}^{\mathrm{c}}$ on $\tilde{\mathcal{W}}^{\mathrm{c}}$ such that

- trajectories of $\tilde{\Phi}^{\mathrm{c}}$ on $\tilde{\mathcal{W}}^{\mathrm{c}}$ are solutions of (50);
- there is a $\delta > 0$ such that for any solution $\underline{u}(r)$, $r \in \mathbb{R}^{+}$ of (44) with $|\underline{u}(r)| < \delta$ for all $r \in \mathbb{R}^{+}$, we have $\underline{u}(r) \in \mathcal{W}^{\mathrm{c}}(r)$ for all $r \in \mathbb{R}^{+}$ and $(\underline{u}(r), r)$, $r \in \mathbb{R}^{+}$ is a trajectory of the flow $\tilde{\Phi}^{\mathrm{c}}$.

THEOREM 3.3. *Assume Hypotheses 2.5 and 2.8. For each* $0 < m < \infty$ *and each* μ *in a small neighborhood of zero, there exists a* C^m*-center manifold* $\tilde{\mathcal{W}}^{\mathrm{c}}$ *for the radial dynamics (44), of dimension* $\dim E^{\mathrm{c}}_{+}/2 + 1$. *The manifold depends* C^m *on the parameter* μ.

The proof will occupy the remainder of this section. We first outline the proof. A modification of the nonlinearity allows us to globalize the problem and consider *all*, not only small, bounded solutions. We then construct two manifolds, which contain solutions bounded for $r > 1$ and for $r < 1$, respectively. The first manifold, $\tilde{\mathcal{W}}^{\mathrm{cs}}_+$, is a global center-stable manifold of equation (49), which contains all solutions with mild exponential growth as $r \to \infty$. The other manifold, $\tilde{\mathcal{W}}^{\mathrm{cu}}_-$, is a global center-unstable manifold of equation (49), which contains all solutions with mild exponential growth as $\tau = \log r \to -\infty$. We show that solutions within $\tilde{\mathcal{W}}^{\mathrm{cu}}_-$ are actually bounded as $\tau \to -\infty$. The intersection, $\tilde{\mathcal{W}}^{\mathrm{cs}}_+ \cap \tilde{\mathcal{W}}^{\mathrm{cu}}_- =: \tilde{\mathcal{W}}^{\mathrm{c}}_{\mathrm{glob}}$, restricted to a neighborhood of $\underline{u} = 0$, gives us the desired, locally invariant center manifold; see Figure 1.

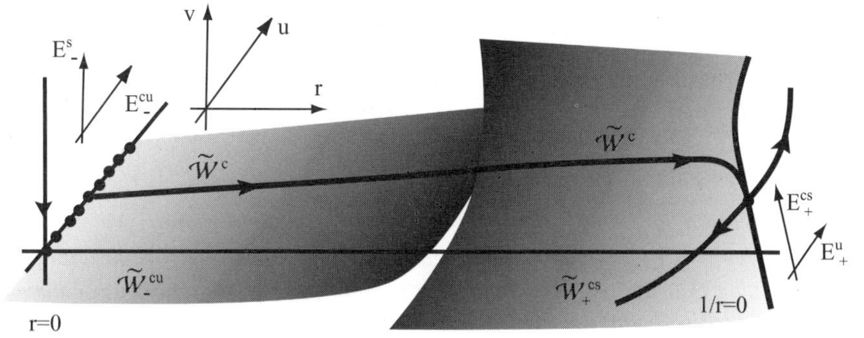

FIGURE 1. *A sketch of the invariant manifolds, needed for the construction of* $\tilde{\mathcal{W}}^{\mathrm{c}}$, *in the extended phase space.*

We first modify F. Let $\chi : \mathbb{R} \to \mathbb{R}$ be a smooth cut-off function with $\chi(s) = 1$ for $s \le 1$, $\chi(s) = 0$ for $s \ge 2$, and $\chi' < 0$ on $(1, 2)$. Define

$$F_{\mathrm{mod}}(u; \mu) := \chi\left(\frac{|u|^2 + |\mu|^2}{\delta'}\right)(F(u; \mu) - \partial_U F(0; 0)u),$$

the δ'-dependent modified nonlinearity. Note that the Lipschitz constant of the modified nonlinearity $F_{\mathrm{mod}}(\cdot; \mu)$ tends to zero

$$\mathrm{Lip}\, F_{\mathrm{mod}}(\cdot; \mu) \to 0, \quad \text{for } \delta' \to 0,$$

uniformly in μ.

We then consider the system (45), with $\mathcal{F}_{\mathrm{mod}}((u, w); \mu) := (0, D^{-1} F_{\mathrm{mod}}(u; \mu))$

$$\underline{u}_r = \mathcal{A}(r)\underline{u} + \mathcal{F}_{\mathrm{mod}}(\underline{u}; \mu). \tag{54}$$

We start constructing $\tilde{\mathcal{W}}^{\mathrm{cs}}_+$. Consider (54) in the extended phase space

$$u' = \mathcal{A}(1/\alpha)\underline{u} + \mathcal{F}_{\mathrm{mod}}(\underline{u}; \mu), \quad \alpha' = -\alpha^2. \tag{55}$$

Note that, by definition, $\mathcal{A}(1/\alpha)$ is smooth in α at $\alpha = 0$.

PROPOSITION 3.4. *For μ in a small neighborhood of zero, equation (55) possesses a flow-invariant center-stable manifold $\tilde{\mathcal{W}}^{\mathrm{cs}}_+$, that contains all solutions which are bounded as $r \to \infty$. The manifold is C^m and depends C^m on μ. Moreover, $\tilde{\mathcal{W}}^{\mathrm{cs}}_+$ is tangent to $(E^{\mathrm{c}}_+ \oplus E^{\mathrm{s}}_+) \times \mathbb{R}$ at $\underline{u} = 0, \alpha = 0$, $\mu = 0$ in the extended phase space $\mathbb{R}^{2N} \times \mathbb{R}$. For any $\alpha_0 < \infty$, $\mathbb{R}^{2N} \times \{|\alpha| \le \alpha_0\} \cap \tilde{\mathcal{W}}^{\mathrm{cs}}_+$ is δ'-close to $(E^{\mathrm{c}}_+ \oplus E^{\mathrm{s}}_+) \times \{|\alpha| \le \alpha_0\}$ in the C^1-topology.*

Proof. We construct the center-stable manifold to (55) near $\alpha = 0, \underline{u} = 0$. Modifying the α-dependent nonlinearities by multiplying them with $\chi(|\alpha|/\delta')$, we obtain a new modified equation which is a small perturbation of the linearization in $\underline{u} = 0$, $\alpha = 0$. We can apply results in [**Van89**] to find a global center-stable manifold $\tilde{\mathcal{W}}^{\mathrm{cs}}_{+,\mathrm{glob}}$. Note that the modification of the α-terms only appears for $\alpha > \delta'$, such that $\tilde{\mathcal{W}}^{\mathrm{cs}}_{+,\mathrm{glob}} \cap \{\alpha < \delta'\}$ is forward invariant under the flow to (55). Transporting this forward invariant manifold backwards with the flow to (55) gives the desired invariant center-stable manifold. ∎

The next step is the construction of $\tilde{\mathcal{W}}^{\mathrm{cu}}_{-}$. We consider the modified version of (50)

$$\dot{\underline{u}} = r\mathcal{A}(r)\underline{u} + r\mathcal{F}_{\mathrm{mod}}(\underline{u};\mu), \qquad \dot{r} = r. \tag{56}$$

PROPOSITION 3.5. *For all μ, not necessarily small, there exists a unique C^∞-center-unstable manifold $\tilde{\mathcal{W}}^{\mathrm{cu}}_{-} \subset \mathbb{R}^{2N} \times \overline{\mathbb{R}^{+}}$, consisting of all solutions to (56) which are bounded as $\tau \to -\infty$. The manifold depends C^∞ on μ. The manifold $\tilde{\mathcal{W}}^{\mathrm{cu}}_{-}$ is δ'-close to $\tilde{E}^{\mathrm{cu}}_{-} = (E^{\mathrm{cu}}_{-} \times \{0\}) \cup \bigcup_{r\le 1} E^{\mathrm{cu}}_{-}(r) \times \{r\}$ on $r \le 1$ in the C^1-topology, and is tangent to $\tilde{E}^{\mathrm{cu}}_{-}$ in $\underline{u} = 0$, $r = 0$ and $\mu = 0$. Moreover, $\tilde{\mathcal{W}}^{\mathrm{cu}}_{-}$ is a C^∞-fiber bundle $\tilde{\mathcal{W}}^{\mathrm{cu}}_{-} = \bigcup_{u_0\in\mathbb{R}^N} \tilde{\mathcal{W}}^{\mathrm{u}}_{-}(u_0)$, where $\tilde{\mathcal{W}}^{\mathrm{u}}_{-}(u_0)$ are the C^∞-, one-dimensional strong unstable manifolds of the equilibria $u = u_0$, $w = 0$, $r = 0$, which consist of a unique single trajectory $((u,w)(\tau), \mathrm{e}^\tau)$ with $(u,w)(\tau) - (u_0, 0) = \mathrm{O}(\mathrm{e}^\tau)$ as $\tau \to -\infty$.*

Proof. The linearization at an equilibrium $u = u_0$, $w = 0$, $r = 0$ is

$$\dot{u} = 0, \quad \dot{w} = -(n-1)w - rD^{-1}F_{\mathrm{mod}}(u_0;0), \quad \dot{r} = r$$

with one-dimensional strong unstable direction and eigenvalue one. The strong unstable manifold, $\tilde{\mathcal{W}}^{\mathrm{u}}_{-}(u_0)$ is a C^∞-manifold, which depends smoothly on the base point u_0. The union of the $\tilde{\mathcal{W}}^{\mathrm{u}}_{-}(u_0)$ form a smooth fiber bundle which is the (unique) center-unstable manifold. The construction is smoothly depending on μ. ∎

Next, we define $\tilde{\mathcal{W}}^{\mathrm{c}}_{\mathrm{glob}} = \tilde{\mathcal{W}}^{\mathrm{cu}}_{-} \cap \tilde{\mathcal{W}}^{\mathrm{cs}}_{+}$, the global center manifold. As an intersection of invariant manifolds, $\tilde{\mathcal{W}}^{\mathrm{c}}_{\mathrm{glob}}$ is invariant. By construction, $\tilde{\mathcal{W}}^{\mathrm{c}}_{\mathrm{glob}}$ contains all bounded solutions of (44). We show that the intersection is a smooth manifold of dimension $\dim E^{\mathrm{c}}_{+}/2$.

PROPOSITION 3.6. *For all μ in a small neighborhood of zero, the set $\tilde{\mathcal{W}}^{\mathrm{c}}_{\mathrm{glob}} = \tilde{\mathcal{W}}^{\mathrm{cu}}_{-} \cap \tilde{\mathcal{W}}^{\mathrm{cs}}_{+}$ is a C^m-manifold of dimension $\dim E^{\mathrm{c}}_{+}/2$, depending C^m on μ.*

Proof. We have to show that $(\mathcal{W}^{\mathrm{c}}_{\mathrm{glob}}(r_0), r_0) = \tilde{\mathcal{W}}^{\mathrm{c}}_{\mathrm{glob}} \cap (Y \times \{r = r_0\})$ is a smooth manifold, which depends smoothly on r_0. We therefore show that the fiber-wise intersection $\mathcal{W}^{\mathrm{cu}}_{-}(r_0) \cap \mathcal{W}^{\mathrm{cs}}_{+}(r_0)$ is transverse in $\underline{u} = 0$. Since both manifolds are δ'-close in the C^m-topology to their tangent space in $\underline{u} = 0$, see Propositions 3.4 and 3.5, we only have to show that their tangent spaces intersect transversely for all $r_0 > 0$. In other words, we have to show that $E^{\mathrm{cu}}_{-}(r_0)$ and $E^{\mathrm{c}}_{+}(r_0) \oplus E^{\mathrm{s}}_{+}(r_0)$ (see Lemma 3.1 for their definition) intersect transversely along $E^{\mathrm{c}}(r_0)$, a linear subspace of dimension $\dim E^{\mathrm{c}}_{+}/2$. The linear spaces are transported by the linearized equation (46). Note first that, if $E \le \mathbb{R}^N$ is invariant under $D^{-1}\partial_U F(0;0)$, then $E \times E \le \mathbb{R}^N \times \mathbb{R}^N$ is invariant under the evolution of (46). We may therefore restrict to showing transversality in flow-invariant subspaces, where $D^{-1}\partial_U F(0;0)$

can be assumed to be in Jordan normal form. Consider the system for a single Jordan block

$$v_r = w, \quad w_r = -\frac{n-1}{r}w - \mathcal{N}^j_\nu v, \tag{57}$$

with $(v, w) \in \mathbb{R}^{2j} = E \times E$ and $\mathcal{N}^j_\nu = \nu\mathrm{id} + \mathcal{N}^j$ and $\mathcal{N}^j$ the standard nilpotent Jordan block of dimension j in E. We claim that within $E \times E$ the intersection of $E^{\mathrm{cu}}_-(r_0)$ and $E^{\mathrm{cs}}_+(r_0)$ is transverse. Let first $\nu \in \mathbb{R}^-$. Then $E\times E \le E^{\mathrm{c}}_+ \le E^{\mathrm{cs}}_+$, which implies transversality. Also, within $E \times E$, we get $E^{\mathrm{c}}(r_0) = E^{\mathrm{cu}}_-(r_0) \cap E^{\mathrm{cs}}_+(r_0) = E^{\mathrm{cu}}_-(r_0)$, which is j-dimensional, half the dimension of $E^{\mathrm{c}}_+ \cap (E \times E)$.

Let $\nu \notin \mathbb{R}^-$, next. Then (57) in $r = \infty$ is hyperbolic and E^{c}_+ is trivial. Thereby, $E^{\mathrm{c}}(r_0) \cap (E \times E) = \{0\}$ and $\dim E^{\mathrm{cs}}_+(r_0) = \dim E^{\mathrm{s}}_+(r_0) = j$. We have to show $E^{\mathrm{cu}}_-(r_0) \cap E^{\mathrm{s}}_+(r_0) = \emptyset$. Note that, by Lemma 3.1, solutions in the intersection give bounded solutions to the linear equation (57). Write $v = (v^1, \ldots, v^j)$. Then v^j solves Bessel's equation

$$v^j_r = w^j, \quad w^j_r = -\frac{n-1}{r}w^j - \nu v^j.$$

For $\nu \notin \mathbb{R}^-$, there are no bounded solutions to this equation [**Wat22**]. Hence, for $(v, w) \in E^{\mathrm{c}}(r_0) \cap (E \times E)$, $v^j = w^j = 0$. The equation for v^i, $i = 1, \ldots, j-1$, is

$$v^i_r = w^i, \quad w^i_r = -\frac{n-1}{r}w^i - \nu v^i - v^{i+1}.$$

If $v^{i+1} \equiv 0$, the above argument shows that also $v^i \equiv 0$, which proves transversality of the intersection $E^{\mathrm{cu}}_-(r_0) \cap (E^{\mathrm{c}}_+(r_0) \oplus E^{\mathrm{s}}_+(r_0))$ by induction on j. Given transversality of the manifolds, Lyapunov-Schmidt reduction proves smoothness of the intersection manifold $\mathcal{W}^{\mathrm{c}}_{\mathrm{glob}}(r_0) = \mathcal{W}^{\mathrm{cu}}_-(r_0) \cap \mathcal{W}^{\mathrm{cs}}_+(r_0)$. Transporting the smooth manifold with the flow gives the r-dependent, smooth manifold $\tilde{\mathcal{W}}^{\mathrm{c}}_{\mathrm{glob}}$. Transversality persists for nearby parameter values μ, which gives us a μ-dependent, smooth family of center manifolds. This proves the proposition. ∎

Proof. *[Theorem 3.3]* Intersecting $\tilde{\mathcal{W}}^{\mathrm{c}}_{\mathrm{glob}}$ with $|\underline{u}| \le \delta'$ gives a center manifold $\tilde{\mathcal{W}}^{\mathrm{c}}$. By construction, $\tilde{\mathcal{W}}^{\mathrm{c}}$ is locally invariant as the intersection of a globally invariant and a locally invariant manifold. It contains all small bounded solutions of the original equation (45), since these are bounded solutions to the modified equation (56), and therefore contained in $\tilde{\mathcal{W}}^{\mathrm{c}}_{\mathrm{glob}}$. This proves Theorem 3.3. ∎

We remark that in case of a stationary, homogeneous instability (O), the center manifold is one-dimensional, $\dim \mathcal{W}^{\mathrm{c}}(r) = 1$. In the Turing case (T), $\dim \mathcal{W}^{\mathrm{c}}(r) = 2$. In both cases, the tangent spaces $E^{\mathrm{c}}(r)$ depend on r and computing Taylor jets of a reduced equation does not seem feasible. Also, the behavior of $\mathcal{W}^{\mathrm{c}}(r)$ for $r \to \infty$ may be very complicated.

2.2. The asymptotic center manifold $\tilde{\mathcal{W}}^{\mathrm{c}}_+$. We construct a larger manifold, $\tilde{\mathcal{W}}^{\mathrm{c}}_+$, which still contains all bounded solutions, but where the tangent space does not depend on r. In addition, this larger manifold possesses a well-defined asymptotic as $r \to \infty$.

THEOREM 3.7. *For all μ sufficiently close to zero, there exists a C^m-center manifold $\tilde{\mathcal{W}}^{\mathrm{c}}_+$ (in the sense of Definition 2.13) for equation (49) near the equilibrium $\underline{u} = 0$, $\alpha = 0$ with local flow $\tilde{\Phi}^{\mathrm{c}}_+$. The center manifold depends C^m on μ and is tangent to $E^{\mathrm{c}}_+ \times \mathbb{R}$ in the extended phase space $(\underline{u}, \alpha) \in \mathbb{R}^{2N} \times \mathbb{R}$, at $\underline{u} = 0$, $\alpha = 0$,* *for* $\mu = 0$. *Moreover,* $\tilde{\mathcal{W}}^{\mathrm{c}}_{+}$ *contains all small bounded solutions* $\underline{u}(r)$, $r \in \mathbb{R}^{+}$*: there are* $\delta' > 0$ *and* $\alpha_0 > 0$ *such that, if* $|\underline{u}(r)| \leq \delta'$ *for all* $r > 0$*, then* $(\underline{u}(r), 1/r) \in \tilde{\mathcal{W}}^{\mathrm{c}}_{+}$ *for all* $r \geq 1/\alpha_0$.

We emphasize that the term center manifold in this theorem refers to Definition 2.13 of an autonomous center manifold and not to Definition 3.2 for the nonautonomous radial dynamics. Figure 2 illustrates the enlarged geometry near $\alpha = 0$.

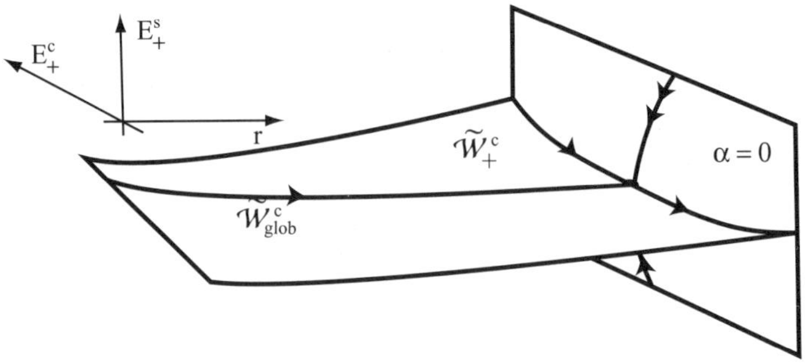

FIGURE 2. *The center manifold* $\tilde{\mathcal{W}}^{\mathrm{c}}_{+}$ *inside* $\tilde{\mathcal{W}}^{\mathrm{c}}_{\mathrm{glob}}$. *The three-dimensional picture is within a neighborhood of* $\underline{u} = 0$, $\alpha = 0$, *inside* $\tilde{\mathcal{W}}^{\mathrm{cs}}_{+}$.

Proof. Existence of a center manifold is a standard result, see for example [**Shu87, Van89**]. Center manifolds are not unique. We want to choose a center manifold, which contains the bounded solutions, $\tilde{\mathcal{W}}^{\mathrm{c}}$. Therefore, consider once again the modified equation with small nonlinearity $\mathcal{F}_{\mathrm{mod}}$ and the global center-stable manifold $\tilde{\mathcal{W}}^{\mathrm{cs}}_{+}$; see Proposition 3.4. Within $\tilde{\mathcal{W}}^{\mathrm{cs}}_{+,\mathrm{glob}} \cap (\mathbb{R}^{2N} \times \{\alpha \leq \alpha_0\})$, we construct a center (-unstable) manifold $\tilde{\mathcal{W}}^{\mathrm{c}}_{+,\mathrm{glob}}$, which is forward invariant and contains $\tilde{\mathcal{W}}^{\mathrm{c}}_{\mathrm{glob}} \cap (\mathbb{R}^{2N} \times \{\alpha = \alpha_0\}) = \mathcal{W}^{\mathrm{c}}_{\mathrm{glob}}(1/\alpha_0)$. Then we restrict $\tilde{\mathcal{W}}^{\mathrm{c}}_{+,\mathrm{glob}}$ to $|\underline{u}| < \delta'$ to obtain the asymptotic center manifold $\tilde{\mathcal{W}}^{\mathrm{c}}_{+}$. To construct $\tilde{\mathcal{W}}^{\mathrm{c}}_{+,\mathrm{glob}}$, we use graph transform as exposed in [**Shu87, KH95**], for example.

The analysis is performed within $\tilde{\mathcal{W}}^{\mathrm{cs}}_{+}$. Dynamics within $\tilde{\mathcal{W}}^{\mathrm{cs}}_{+}$ are determined by the projection on $E^{\mathrm{cs}}_{+} = E^{\mathrm{c}}_{+} \oplus E^{\mathrm{s}}_{+}$. We therefore consider only graphs within E^{cs}_{+}, transported with the flow on $\tilde{\mathcal{W}}^{\mathrm{cs}}_{+}$, projected on E^{cs}_{+}. Moreover, we restrict to graphs over $\mathcal{U} = E^{\mathrm{c}}_{+} \times \{|\alpha| \leq \alpha_0\}$, which can be identified with maps from $\mathcal{U}$ into E^{s}_{+}. Define

$$\Sigma_\delta = \left\{\psi \in BC^0(\mathcal{U}, E^{\mathrm{s}}_{+}); \sup|\psi| \leq \delta, \operatorname{Lip}\psi \leq 1\right\}.$$

Together with the distance, induced by the sup-norm on $BC^0(\mathcal{U}, E^{\mathrm{s}}_{+})$, Σ_δ becomes a closed metric space. We choose and fix $\mathcal{W}^{\mathrm{c}}_{+,\mathrm{glob}}(\alpha_0) = \tilde{\mathcal{W}}^{\mathrm{c}}_{+,\mathrm{glob}} \cap (\mathbb{R}^{2N} \times \{\alpha = \alpha_0\})$ as an (arbitrary) extension of $\mathcal{W}^{\mathrm{c}}_{\mathrm{glob}}(1/\alpha_0)$, tangent to $E^{\mathrm{c}}_{+}(1/\alpha_0)$. For example, we write $E^{\mathrm{c}}(r_0)$ for the tangent space of $\mathcal{W}^{\mathrm{c}}_{\mathrm{glob}}(r_0)$ in $(u, v) = 0$ and decompose $E^{\mathrm{c}}_{+}(r_0) = E^{\mathrm{c}}(r_0) \oplus C$ with arbitrary complement C. We then define

$$\mathcal{W}^{\mathrm{c}}_{+,\mathrm{glob}}(1/\alpha_0) = \left\{(\psi^{\mathrm{c}}_{0,\mathrm{glob}}(u^{\mathrm{c}}) + u^{\mathrm{c}} + v); u^{\mathrm{c}} \in E^{\mathrm{c}}(r_0), v \in C\right\},$$

where $\mathcal{W}^{\mathrm{c}}_{\mathrm{glob}}(1/\alpha_0) = \operatorname{graph} \psi^{\mathrm{c}}_{0,\mathrm{glob}}$. We restrict Σ_δ to graphs respecting the boundary value:

$$\Sigma^*_\delta = \left\{\psi \in \Sigma_\delta; \operatorname{graph} \psi(\alpha_0, \cdot) = (\mathcal{W}^{\mathrm{c}}_{+,\mathrm{glob}}(1/\alpha_0), \alpha_0)\right\}.$$

With the topology induced from Σ_δ, Σ^*_δ is again a closed metric space.

We define the graph transform next. Let $\tilde{\Phi}_r$ denote the flow in $\tilde{\mathcal{W}}^{\mathrm{cs}}_+$, projected on E^{cs}_+ and let $\tilde{\Phi}_1$ be the time-one map. We define the graph transform $\mathcal{T}\psi$ as a map on $\psi \in \Sigma^*_\delta$ by

$$\operatorname{graph}(\mathcal{T}\psi) = \tilde{\Phi}_1(\operatorname{graph}\psi) \cup \mathcal{M}_{\mathrm{in}},$$

and

$$\mathcal{M}_{\mathrm{in}} := \bigcup_{0 \le r \le 1} \tilde{\Phi}_r((\alpha_0, \mathcal{W}^{\mathrm{c}}_{+,\mathrm{glob}}(\alpha_0))),$$

where $\mathcal{W}^{\mathrm{c}}_{+,\mathrm{glob}}(\alpha_0)$ is the boundary manifold. It is now tedious but straightforward, following for example [**Shu87**], to verify that $\mathcal{T}$ is a well-defined map on Σ^*_δ. Also, a sufficiently high iterate $\mathcal{T}^j$ is a contraction, due to exponential contraction of the linearized flow in the direction E^{s}_+, where the graphs take values. The unique fixed point is forward invariant under $\tilde{\Phi}_1$ by construction, and invariant under $\tilde{\Phi}_r$ for all $r > 0$ by uniqueness, see again [**Shu87**]. Defining $\tilde{\mathcal{W}}^{\mathrm{c}}_+$ as the unique fixed point, intersected with a small neighborhood of $(u, v) = 0$, we have obtained the desired center manifold. ∎

Transporting $\tilde{\mathcal{W}}^{\mathrm{c}}_{+,\mathrm{glob}}$ backwards with the flow $\tilde{\Phi}$, we can construct a flow-invariant manifold (which is in addition backward invariant), although we do not need this construction later on. For the non modified, original vector field (44), this globalization of $\tilde{\mathcal{W}}^{\mathrm{c}}_{+,\mathrm{glob}}$ restricted to a small neighborhood of $\underline{u} = 0$ is a center manifold in the sense of Definition 3.2, but of twice the dimension as $\tilde{\mathcal{W}}^{\mathrm{c}}$ in each fiber.

REMARK 3.8. *Reflecting the construction at $\alpha = 0$, allows for a construction of $\tilde{\mathcal{W}}^{\mathrm{c}}_+$ also for negative radii, on $|\alpha| \le \alpha_0$. The center manifold is then invariant under the reflection $R : ((u, v), \alpha) \mapsto ((u, -v), -\alpha)$. The flow Φ^{c}_+ is reversible with respect to R.*

3. Expansions and normal forms

The goal of this section is to derive expansions for the bifurcation equations on $\tilde{\mathcal{W}}^{\mathrm{c}}_+$. We start with the linearized equation in Section 3.1, identifying the linear nonautonomous equation in E^{c}_+. In case of a Turing instability, the dependence on r can be simplified using smooth, r-dependent, linear transformations. In Section 3.2, we then use the same strategy to simplify higher-order terms in the spirit of the normal form theory in Chapter 2, Section 3.3.

Since the tangent space and then the Taylor jet of $\tilde{\mathcal{W}}^{\mathrm{c}}_+$ in $(u, v, \alpha) = 0$ is uniquely determined, the particular construction of $\tilde{\mathcal{W}}^{\mathrm{c}}_+$ and the matching manifold $\tilde{\mathcal{W}}^{\mathrm{cu}}_-$ do not play a role in this section.

We restrict to the case of $n = 2$ for the rest of this chapter. Most of the arguments are identical for the case $n > 2$.

3.1. The linearized equation and expansions. The Taylor jet of the reduced vector field on $\tilde{\mathcal{W}}^{\mathrm{c}}_{+}$ is uniquely determined and does not depend on the particular construction involving $\tilde{\mathcal{W}}^{\mathrm{cu}}_{-}$ in the proof of Theorem 3.7. We expand the center manifold and the reduced vector field on $E^{\mathrm{c}}_{+} \times \{|\alpha| \le \alpha_0\}$ in the variable $\underline{u}^{\mathrm{c}} \in E^{\mathrm{c}}_{+}$ and α. Note that in the invariant subspace $\alpha = 0$ we obtain the reduced equations from Chapter 2, equations (14), in case of a homogeneous instability, and (15), in case of a Turing instability.

LEMMA 3.9. *In case of a nondegenerate stationary homogeneous instability (O), the reduced vector field on* $E^{\mathrm{c}}_{+} \times \{|\alpha| \le \alpha_0\}$ *takes the form*

$$
(58)\qquad
\begin{aligned}
A' &= B\\
B' &= -\alpha B + \gamma_0(\mu,\alpha) + \gamma_1(\mu,\alpha)A + \gamma_2(\mu,\alpha)A^2 + \gamma_3(\mu,\alpha)A^3\\
&\quad +\mathcal{R}(A,B,\alpha;\mu)\\
\alpha' &= -\alpha^2,
\end{aligned}
$$

in the coordinates from (14). The coefficients γ_j *satisfy*

$$
(59)\qquad \gamma_0(0,\alpha) = 0,\ \gamma_1(0,\alpha) = 0,\ \partial_\mu\gamma_0(0,0) \ne 0,\ \text{and } \gamma_2(0,\alpha) \equiv \gamma_2(0,0).
$$

The remainder terms are

$$
\mathcal{R} = \mathrm{O}\left((|\mu| + |\alpha|^2)|\alpha B| + |\alpha AB| + |B|^3 + \sum_{j=0}^{2} |A^j B^{3-j}| + (|A| + |B|)^4 \right).
$$

The full equation is reversible with respect to

$$
(60)\qquad (A,B,\alpha,r) \mapsto (A,-B,-\alpha,-r).
$$

Proof. We expand the equation for α small and set $A' =: B$. We have to check the conditions on the coefficients γ_j, (59) and verify the structure of the remainder $\mathcal{R}$.

For $\alpha = 0$, the expansions for the vector field and the remainder $\mathcal{R}$ are as derived in Chapter 2. For $\alpha \ne 0$, the coefficients depend on α with the only restriction imposed by reversibility, (60). Reversibility of the equation on the center manifold was already observed in Remark 3.8.

We verify (59), next. For $\mu = 0$, $\underline{u} = 0$ is a solution and therefore, $\gamma_0(0,\alpha) = 0$. In $\alpha = 0$, we have $\partial_\mu\gamma_0 \ne 0$, which are the conditions on γ_0 claimed in (59).

It remains to show that $\gamma_1(0,\alpha) = 0$ and $\gamma_2(0,\alpha) \equiv \gamma_2(0,0)$. For $\mu = 0$, the linear part of the reduced vector field is the restriction of the full linearized vector field (46) to $E^{\mathrm{c}}_{+}(r)$. Now E^{c}_{+} is spanned by $(U_0, 0)$ and $(0, U_0)$ with U_0 spanning the kernel of $\partial_U F(0;0)$. For the nonautonomous equation

$$
v' = w, \qquad w' = -\frac{1}{r}w - D^{-1}\partial_U F(0;0)v,
$$

this center space is invariant: $E^{\mathrm{c}}_{+}(r) \equiv E^{\mathrm{c}}_{+}$. Restricting to this invariant subspace gives $A' = B$, $B' = -\alpha B$, as claimed. The linear terms in μ as well as the quadratic terms in A are obtained by projecting the α-independent part of the vector-field on the α-independent linear center subspace and are therefore α-independent as well, which shows that $\gamma_2(0,\alpha)$ is in fact independent of α. ∎

In case of a Turing instability, we need normal form transformations. We adapt the strategy from Theorem 2.17 on nonsemisimple normal forms to the nonautonomous setting, here. We start analyzing the linear part.

LEMMA 3.10. *Consider a linearly generically unfolded Turing instability. Then for any $0 < m < \infty$ the reduced equation on the center eigenspace E^{c}_{+} can be written in the following form:*

$$
\begin{aligned}
(61)\ A' &= (\mathrm{i}k(\alpha;\mu) + \nu(\alpha;\mu))A + B + \mathrm{O}\left((|A|+|B|)^2 + |\mu||\alpha|^m(|A|+|B|)\right)\\
B' &= \gamma_1(\alpha;\mu)A + (\mathrm{i}k(\alpha;\mu) + \nu(\alpha;\mu))B\\
&\quad +\mathrm{O}\left((|A|+|B|)^2 + |\mu||\alpha|^m(|A|+|B|)\right)\\
\alpha' &= -\alpha^2
\end{aligned}
$$

with $\partial_\mu\gamma_1(0;0) \neq 0$, $k(\alpha;0) = \mathrm{O}(|\alpha|^3)$, $\nu(\alpha;0) = -\frac{\alpha}{2} + \mathrm{O}(|\alpha|^3)$, and $\gamma_1(\alpha;0) = \mathrm{O}(|\alpha|^3)$. Here, A and B are defined as for the one-dimensional problem, up to order α; see Lemma 2.11,.

Proof. We first show that, again, $E^{\mathrm{c}}_{+}(r) \equiv E^{\mathrm{c}}_{+}$. In other words, the solutions to the linear equation (46) which neither grow nor decay exponentially as $r \to \infty$ span an r-independent linear subspace of $Y = \mathbb{R}^{2N}$. Recall that $\mathrm{spec}\,(-D^{-1}\partial_U F(0;0)) \cap \mathbb{R}^- = \{-k_*^2\}$, geometrically simple and algebraically double. We may transform $D^{-1}\partial_U F(0;0)$ into Jordan normal form conjugating with T. The diagonal transformation $(u,w) = (T\tilde{u}, T\tilde{w})$ puts the linearization $\mathcal{A}(\infty)$ into block-diagonal structure, which shows invariance of E^{c}_{+}. On E^{c}_{+}, we obtain the restriction

$$
\mathcal{A}^{\mathrm{c}}(\alpha) = \begin{pmatrix} 0 & 0 & 1 & 0\\ 0 & 0 & 0 & 1\\ -k_*^2 & 1 & -\alpha & 0\\ 0 & -k_*^2 & 0 & -\alpha \end{pmatrix}.
$$

For convenience, we write $\mathcal{A}^{\mathrm{c}}(\alpha)$ as a function of α instead of r, since we want to use transformations which are smooth in the variable $\alpha = 1/r$. Set $\underline{u}^{\mathrm{c}} = A(1,0,\mathrm{i}k_*,0)^T + B(0,2\mathrm{i}k_*,1,-2k_*^2)$+c.c., with reversibility acting through $(A,B) \mapsto (\overline{A}, -\overline{B})$ and $\alpha \mapsto -\alpha$. We obtain

$$
\begin{aligned}
(62)\qquad A' &= (\mathrm{i}k_* - \frac{\alpha}{2})A + B + \frac{\alpha}{2}\overline{A} + \mathrm{O}\left((|\mu| + |A| + |B|)(|A|+|B|)\right)\\
B' &= (\mathrm{i}k_* - \frac{\alpha}{2})B - \frac{\alpha}{2}\overline{B} + \mathrm{O}\left((|\mu| + |A| + |B|)(|A|+|B|)\right)\\
\alpha' &= -\alpha^2.
\end{aligned}
$$

Estimates for the higher-order terms are uniform in α small.

Next, we use normal form transformations to eliminate dependence on $\overline{A}$ and $\overline{B}$ on the right side. We subsequently transform $\underline{u}^{\mathrm{c}}_{j+1} = \underline{u}^{\mathrm{c}}_j + \alpha^j T_j \underline{u}^{\mathrm{c}}_j$, $\underline{u}^{\mathrm{c}}_1 = \underline{u}^{\mathrm{c}}$, $j = 1, \ldots, m-1$, with suitable linear, invertible operators T_j. If $(\underline{u}^{\mathrm{c}}_j)' = \mathcal{A}_j(\alpha)\underline{u}^{\mathrm{c}}_j$, then

$$
(63)\qquad (\underline{u}^{\mathrm{c}}_{j+1})' = \mathcal{A}_j(\alpha)\underline{u}^{\mathrm{c}}_{j+1} + \alpha^j(T_j\mathcal{A}^{\mathrm{c}}(0) - \mathcal{A}^{\mathrm{c}}(0)T_j)\underline{u}^{\mathrm{c}}_{j+1} + \mathrm{O}\left(|\alpha|^{j+1}\right).
$$

We may therefore subsequently eliminate all terms which are not part of the Arnol'd unfolding of $\mathcal{A}(0)$, up to any prescribed, finite order in α. We obtain the Arnol'd normal form as stated in (61) for $\mu \neq 0$..

The transformations T_j can be computed explicitly. Note that the first transformation T_1 eliminates the α-dependent terms involving $\overline{A}$ and $\overline{B}$ in the equations for A and B.

As a next step, we show that we may indeed eliminate α-dependence in the linear part completely at $\mu = 0$, up to the terms γ_1, ν, and $\mathrm{i}k$, explicited in (61).

To start with, set $\tilde{B} = B + s\overline{B}$ with $s = s(\alpha)$ some function of α to be determined later. From (61), abbreviating $\lambda = \mathrm{i}k_* - \frac{\alpha}{2}$, we then obtain

$$\tilde{B}' = \tilde{B}\frac{1}{1-s\bar{s}}(\lambda - s\bar{s}\bar{\lambda} - \frac{\alpha}{2}(s-\bar{s}) - \bar{s}s_r)$$

if

$$s' = -2\mathrm{i}k_* s + \alpha\frac{s^2-1}{2}, \quad \alpha' = -\alpha^2.$$

Going into a corotating frame $\tilde{s} = \mathrm{e}^{2\mathrm{i}k_* r}s$, gives

$$\tilde{s}' = \frac{1}{2}(\mathrm{e}^{-2\mathrm{i}k_* r}\tilde{s}^2 - \mathrm{e}^{2\mathrm{i}k_* r})\alpha, \quad \alpha' = -\alpha^2.$$

Rewritten as an integral equation

$$\tilde{s}(r) = \int_\infty^r \frac{\mathrm{e}^{-2\mathrm{i}k_*\rho}\tilde{s}^2(\rho) - \mathrm{e}^{2\mathrm{i}k_*\rho}}{2\rho}\mathrm{d}\rho$$

we see that the right side defines a contraction in the space of functions $\tilde{s}(\rho)$, $\rho \geq R > 0$ with R large, and norm $\sup_{\rho\geq R}|s(\rho)|/\rho$.

Next, transforming $\tilde{A} = A - s\overline{A} + t_1\overline{\tilde{B}}$ with s as above, and a suitable α-dependent coefficient t_1, and rescaling $\tilde{B}$ appropriately gives the desired normal form. The coefficient t_1 is found to leading order as $-s/(\mathrm{i}k_*(1-s\bar{s})) + \mathrm{O}(r^{-2})$.

This shows that the normal form transformations can, for $\mu = 0$, be carried out by a smooth function of α, leaving no remainder terms of the form $\mathrm{O}(|\alpha^m|)$.

It remains to show the expansions for the coefficients k, ν, and γ_1.

The claim on the expansion for γ_1 follows from the considerations for the asymptotic problem, Lemma 2.11, and reversibility of the system. Inspecting the definition of the transformations, we readily derive the linear term in ν and see that quadratic terms in α in the coefficients ν, γ_1, and k vanish at $\mu = 0$. ■

3.2. Nonautonomous normal forms. We adapt the normal form algorithm to the nonautonomous set-up. We concentrate on case (T) and expand the reduced equation on E^{c}_+ as follows, dropping the superscript c of the phase space variable $\underline{u}^{\mathrm{c}}$:

$$\underline{u}' = \mathcal{A}(\alpha;\mu)\underline{u} + \sum_{j=2}^{m}\mathcal{G}_j(\underline{u},\alpha;\mu) + \mathcal{R}(\underline{u},\alpha;\mu), \tag{64}$$

where $\mathcal{A}(\alpha,\mu)$ is the linear part as described in Lemma 3.10, the $\mathcal{G}_j(\cdot,\alpha;\mu)$ are homogeneous polynomials of degree j, and the remainder is $\mathcal{R}(\underline{u},\alpha;\mu) = \mathrm{O}(|\underline{u}|^{m+1})$.

PROPOSITION 3.11. *For every $0 < m < \infty$ and μ small, there is an α-dependent, change of coordinates*

$$\underline{v} = \underline{u} + \Phi(\underline{u},\alpha;\mu),$$

with $\Phi(\underline{u},\alpha;\mu) = \mathrm{O}(|\underline{u}|^2)$, such that the new variable $\underline{v}$ solves the equation in normal form up to order m

$$\underline{v}' = \mathcal{A}(\alpha;\mu)\underline{v} + \sum_{j=2}^{m}\mathcal{N}_j(\underline{v},\alpha;\mu) + \tilde{\mathcal{R}}(\underline{v},\alpha;\mu). \tag{65}$$

The transformation Φ can be chosen polynomial in $\underline{u}$ and α. The remainder satisfies $\tilde{\mathcal{R}}(\underline{v},\alpha;\mu) = \mathrm{O}(|\underline{v}|^{m+1} + |\alpha|^{m+1}|\underline{v}|^2)$ and the $\mathcal{N}_j$ are homogeneous polynomials of *degree j in $\underline{v}$, with coefficients being polynomials in α, satisfying the normal form condition*

$$\mathrm{e}^{\mathcal{A}^*(0;0)\sigma}\mathcal{N}_j(\underline{v},\alpha;\mu) = \mathcal{N}_j(\mathrm{e}^{\mathcal{A}^*(0;0)\sigma}\underline{v},\alpha;\mu),$$

for all $\sigma \in \mathbb{R}$.

Proof. We adapt the proof of Theorem 2.17. Again, we proceed iteratively, searching in each step a transformation $\underline{v} = \underline{u} + \Phi(\underline{u},\alpha;\mu)$, with Φ a homogeneous polynomial of degree ℓ in $\underline{u}$. The transformation should eliminate ℓ-th order terms in the Taylor series (64) that are not in normal form, up to order m in α. To find Φ, we start a sub-iteration, eliminating successively terms of the form $\alpha^j\mathcal{P}$ such that the remaining terms of order ℓ in $\underline{u}$ are in normal form up to order j in α. We therefore use transformations of the form

$$\underline{u} = \underline{v} + \alpha^j\Phi_\ell^j(\underline{v};\mu).$$

Suppose at level $j-1$ of the sub-iteration, the equation is given as

$$(66)\qquad \begin{aligned} \underline{u}' &= \mathcal{A}(\alpha;\mu)\underline{u} + \sum_{i=2}^{\ell-1}\mathcal{N}_i(\underline{u},\alpha;\mu) + \sum_{k=0}^{j-1}\alpha^k\mathcal{N}_\ell^k(\underline{u};\mu) + \alpha^j\mathcal{P}_\ell^j(\underline{u};\mu) \\ &\quad + \mathrm{O}(|\alpha|^m|\underline{u}|^2 + |\underline{u}|^{\ell+1} + |\alpha|^{j+1}|\underline{u}|^\ell) \\ \alpha' &= -\alpha^2. \end{aligned}$$

Here, the $\mathcal{N}_i$ represent the normal form part up to order $\ell-1$. They are polynomials in α and $\underline{u}$, homogeneous of degree i in $\underline{u}$. The polynomials $\mathcal{P}_\ell^j$ and $\mathcal{N}_\ell^k$ are homogeneous of degree ℓ in $\underline{u}$. Then the transformed equation reads

$$(67)\qquad \begin{aligned} \underline{v}' &= \mathcal{A}(\alpha;\mu)\underline{v} + \sum_{i=2}^{\ell-1}\mathcal{N}_i(\underline{v},\alpha;\mu) + \sum_{k=0}^{j-1}\alpha^k\mathcal{N}_\ell^k(\underline{v};\mu) \\ &\quad + \alpha^j(\mathcal{P}_\ell^j(\underline{v};\mu) + \mathcal{A}(0;0)\Phi_\ell^j(\underline{v};\mu) - \partial_{\underline{v}}\Phi_\ell^j(\underline{v};\mu)\mathcal{A}(0;0)\underline{v}) \\ &\quad + \mathrm{O}\left(|\alpha|^m|v|^2 + |v|^{\ell+1} + |\alpha|^{j+1}|v|^\ell\right) \\ \alpha' &= -\alpha^2. \end{aligned}$$

Following the proof of Theorem 2.17, we choose Φ_ℓ^j such that $\mathcal{P}_\ell^j + \mathcal{A}(0;0)\Phi_\ell^j - \partial_{\underline{v}}\Phi_\ell^j\mathcal{A}(0;0)\underline{v}$ to be in normal form. By induction on j, we may therefore remove all non normal-form terms of degree ℓ in $\underline{v}$, up to order m in α. Induction on ℓ then proves the proposition. ∎

REMARK 3.12. *The above result can be considerably refined. After a more careful analysis of the homological equation, one can show that terms of order $|\alpha|^m$ can be eliminated completely for m sufficiently large, depending on the length of Jordan chains in the representation of the linearization $\mathcal{A}(0;0)$ on the homogeneous polynomials. Without expanding in $\alpha = 1/r$, we can treat the homological equation at level ℓ as a differential equation in the radius r, where we look for bounded solutions as $r \to \infty$. The second line in (67), the homological equation, then reads at each fixed order ℓ*

$$\Phi' + \mathrm{ad}_\ell\,\mathcal{A}(\frac{1}{r},\mu)\Phi = \mathrm{O}(r^{-m})$$

with $(\mathrm{ad}_\ell\,\mathcal{A}\Phi)(v) = \mathcal{A}\Phi(v) - \partial_v\Phi(v)\mathcal{A}v$. The eigenvalues of $\mathrm{ad}_\ell\,\mathcal{A}(0,0)$ are sums of eigenvalues of $\mathcal{A}(0,0)$ and they are all located on the imaginary axis. Therefore, the fundamental solution of the left hand side of the inhomogeneous differential equation *grows at most polynomially, with exponent given by the length of the longest Jordan chain in* $\mathrm{ad}_\ell\,\mathcal{A}(0,0)$*, minus one. The differential equation can therefore be solved by separation of variables and the solution is found to possess a formal expansion in* r^{-1}.

REMARK 3.13. *Instead of eliminating terms in the range of* $\mathrm{ad}_\ell\,\mathcal{A}$*, we can first split* $\mathcal{A} = \mathcal{A}_S + \mathcal{A}_N$ *into semi-simple and nilpotent parts and only eliminate terms in the rang of* $\mathrm{ad}_\ell\,\mathcal{A}_S$*. For this semi-simple part, no Jordan-Blocks occur in the adjoint representation and terms can be eliminated completely with* α*-dependence. We may then continue with nonsemisimple normal form transformations, which however might leave some* α*-dependence in the coefficients of the remaining normal form polynomials.*

For a Turing instability, the semi-simple normal form transformations yield a normal form symmetry, acting as complex diagonal rotation as in the one-dimensional case. This normal form symmetry is respected to any order in $\underline{u}$ *for all values of* α*, small.*

We believe that not all terms can be removed and that any normal form contains nonlinear terms which explicitly depend on the radius r.

COROLLARY 3.14. *Consider a Turing instability (T) for* μ *and* α *close to zero, in the coordinates of Proposition 2.16. Fix* $0 < m < \infty$*. Then there is a change of coordinates, polynomial in* (A, B, α)*, smooth in* μ *and* $\mathrm{O}(|\mu| + |\alpha| + |A| + |B|)$*-close to the identity such that the vector field on* $\tilde{\mathcal{W}}^{\mathrm{c}}_{+}$ *is given by*

$$
\begin{aligned}
A' &= (\mathrm{i}k(\alpha;\mu) + \nu(\alpha;\mu))A + B + \mathrm{i}\gamma_3(\alpha;\mu)A|A|^2 \\
&\quad + \mathrm{O}\left(\sum_{j=0}^{2} |A^j B^{3-j}| + (|A| + |B|)^5 + |\mu||\alpha|^m(|A| + |B|)\right) \\
B' &= \gamma_1(\alpha;\mu)A + (\mathrm{i}k(\alpha;\mu) + \nu(\alpha))B + \gamma_2(\alpha;\mu)A|A|^2 \\
&\quad + \gamma_4(\alpha;\mu)A|A|^4 + \mathrm{i}\gamma_3(\alpha;\mu)B|A|^2 \\
&\quad + \mathrm{O}\Bigg(|AB^2| + |B^3| + \sum_{j=0}^{4} |A^j B^{5-j}| + (|A| + |B|)^7 \\
&\quad + |\mu||\alpha|^m(|A| + |B|)\Bigg) \\
\alpha' &= -\alpha^2
\end{aligned}
\tag{68}
$$

Rescaling time, and going to corotating coordinate frame, the equation can be simplified to

$$
\begin{aligned}
A' &= \nu(\alpha;\mu)A + B \\
&\quad + \mathrm{O}\left(\sum_{j=0}^{2} |A^j B^{3-j}| + (|A| + |B|)^5 + |\mu||\alpha|^m(|A| + |B|)\right) \\
B' &= \gamma_1(\alpha;\mu)A + \nu(\alpha)B + \gamma_2(\alpha;\mu)A|A|^2 + \gamma_4(\alpha;\mu)A|A|^4 \\
&\quad + \mathrm{O}\Bigg(|AB^2| + |B^3| + \sum_{j=0}^{4} |A^j B^{5-j}| + (|A| + |B|)^7 \\
&\quad + |\mu||\alpha|^m(|A| + |B|)\Bigg) \\
\alpha' &= -\alpha^2(1 + \mathrm{O}(|A|^2 + |\alpha| + |\mu|))
\end{aligned}
\tag{69}
$$

Proof. The proof is an immediate consequence of Proposition 3.11 and Remark 3.13, using the representation of the normal form in $\alpha = 0$ from Proposition 2.18. ∎

4. Matching and transversality

In this section, we derive our main results. We prove existence and uniqueness (in an appropriate sense) of branches of bifurcating, radially symmetric stationary patterns in the three cases of a fold, a cusp, and a Turing instability. For $r \to \infty$, the patterns converge to an equilibrium in the homogeneous case (O) and to a Turing pattern in case (T). So far, we have derived expansions for the ODE-description of bounded radially symmetric patterns in the far-field, $\alpha = 1/r$ small. We start in Section 4.1 scaling the equations on $\tilde{\mathcal{W}}^{\mathrm{c}}_{+}$ with the bifurcation parameter μ. The scalings are the same as obtained in the one-dimensional case, Chapter 2, Sections 3.4 and 3.5. Since the scalings are in all cases long-wavelength expansions, the finite hole $r \leq r_0 = 1/\alpha_0$, left out by the description on $\tilde{\mathcal{W}}^{\mathrm{c}}_{+}$ becomes very small. In other words, considering backwards dynamics in r on the asymptotic center manifold $\tilde{\mathcal{W}}^{\mathrm{c}}_{+}$, we stay within $r \geq r_0$ for large scaled times and long-time dynamics on $\tilde{\mathcal{W}}^{\mathrm{c}}_{+}$ become relevant. As a second step, we isolate bounded solutions to the scaled equation which would stay bounded if the equation on $\tilde{\mathcal{W}}^{\mathrm{c}}_{+}$ was valid for all $r > 0$, Section 4.2. The last, major step is then a matching procedure in Section 4.3. We compute an expansion for the manifold $\tilde{\mathcal{W}}^{\mathrm{cu}}_{-} \cap \tilde{\mathcal{W}}^{\mathrm{c}}_{+}$ at the boundary of $\tilde{\mathcal{W}}^{\mathrm{c}}_{+}$, $r = r_0$. We then find intersections of this matching manifold with the set of solutions $\tilde{W}^{\mathrm{s}}_{+}$ inside $\tilde{\mathcal{W}}^{\mathrm{c}}_{+}$, converging to a prescribed equilibrium as $r \to \infty$. Main tool is the λ-Lemma, which gives us expansions for the right matching manifold $\tilde{\mathcal{W}}^{\mathrm{c}}_{+}$. As a major difficulty in the case of a Turing instability (T), the averaging effect, made visible with the normal form transformations, fundamentally changes the influence of curvature, represented by the α-dependent terms, in the far-field, $\tilde{\mathcal{W}}^{\mathrm{c}}_{+}$, compared to the core region, $\tilde{\mathcal{W}}^{\mathrm{cu}}_{-}$. Close to the center, normal form transformations break down and produce nonadiabatic locking, compare Chapter 2, Section 3.5. We recover transversality at a very low order, in spite of the averaging effects in the far-field.

4.1. Scaling on $\tilde{\mathcal{W}}^{\mathrm{c}}_{+}$ and universal equations. We consider the homogeneous case (O), first and treat the Turing instability (T) later. Assume that we have a fold in the kinetics, that is, we assume in Lemma 3.9, that $\mu \in \mathbb{R}$, $\partial_\mu \gamma_0(0,0) = 1$, and $\gamma_2(0,0) = -1$; Hypothesis 2.20. The reduced equation on $\tilde{\mathcal{W}}^{\mathrm{c}}_{+}$ then becomes

$$\begin{aligned} A' &= B \\ B' &= -\alpha B + \mu - A^2 + \mathcal{R}(A, B, \alpha; \mu) \\ \alpha' &= -\alpha^2, \end{aligned}$$

with $\mathcal{R}(A, B, \alpha; \mu) = \mathrm{O}(|\mu|^2 + (|A| + |B|)^3 + |\mu|(|A| + |B|))$.

Since for $\mu < 0$ the asymptotic equation in $\alpha = 0$ does not possess small invariant sets, Remark 2.22, we assume $\mu > 0$ in the sequel.

We scale $A = |\mu|^{1/2}\tilde{A}$, $B = |\mu|^{3/4}\tilde{B}$, $r = |\mu|^{-1/4}\tilde{r}$, and $\alpha = \tilde{\alpha}|\mu|^{1/4}$. The scaled equation is

$$
\begin{aligned}
\tilde{A}_{\tilde{r}} &= \tilde{B} \\
\tilde{B}_{\tilde{r}} &= -\tilde{\alpha}\tilde{B} + 1 - \tilde{A}^2 + |\mu|^{1/2}|\tilde{\mathcal{R}} \\
\tilde{\alpha}_{\tilde{r}} &= -\tilde{\alpha}^2
\end{aligned}
\tag{70}
$$

with $\tilde{\mathcal{R}} = \tilde{\mathcal{R}}_j(\tilde{A}, \tilde{B}, |\mu|^{1/4}\tilde{\alpha}; \mu)$ bounded when its arguments is bounded, uniformly in $\tilde{\alpha} \leq |\mu|^{-1/4}\alpha_0$.

We next consider the case of a cusp, Hypothesis 2.23. The quadratic term vanishes at bifurcation $\gamma_2(0,0) = 0$, and we need two parameters $\mu = (\mu_1, \mu_2)$. Assume that $\partial_\mu\gamma_0(0,0) = (1,0)$, $\partial_\mu\gamma_1(0,0) = (0,1)$, and $\gamma_3(0,0) = 1$. We are particularly interested in the case of multi-stability in the kinetics, and we therefore assume $\mu_2 < 0$; see also Remark 2.25. We then scale $\mu_1 = \tilde{\mu_1}|\mu_2|^{3/2}$, $A = |\mu_2|^{1/2}\tilde{A}$, $B = |\mu_2|\tilde{B}$ $r = |\mu_2|^{-1/2}\tilde{r}$, and $\alpha = \tilde{\alpha}|\mu_2|^{1/2}$. The scaled equation is

$$
\begin{aligned}
\tilde{A}_{\tilde{r}} &= \tilde{B} \\
\tilde{B}_{\tilde{r}} &= -\tilde{\alpha}\tilde{B} + \tilde{\mu_1} - \tilde{A} + \tilde{A}^3 + |\mu_2|^{1/2}\tilde{\mathcal{R}}(\tilde{A}, \tilde{B}, |\mu_2|^{1/2}\tilde{\alpha}; \mu) \\
\tilde{\alpha}_{\tilde{r}} &= -\tilde{\alpha}^2
\end{aligned}
\tag{71}
$$

with $\tilde{\mathcal{R}}$ bounded when its arguments are bounded, uniformly in $\tilde{\alpha} \leq |\mu_2|^{-1/2}\alpha_0$.

Consider the case of a Turing instability (T), next. We start from (69), where we assume $\gamma_2(0;0) > 0$, the generic, supercritical case. We need a single parameter $\mu \in \mathbb{R}$. Performing an exact normal form transformation to quadratic order in (A, B), see Remark 3.12 and Remark 2.19, in particular, removing all quadratic terms for all α small, we arrive at

$$
\begin{aligned}
A' &= B - \frac{\alpha}{2}A + \mathrm{O}\left((|A| + |B|)^3\right) \\
B' &= \gamma_1(\mu, \alpha)A - \frac{\alpha}{2}B + \gamma_2(\mu, \alpha)A|A|^2 \\
&\quad + \mathrm{O}\left(\sum_{j=0}^{2} |A^j B^{3-j}| + (|A| + |B|)^5\right) \\
\alpha' &= -\alpha^2.
\end{aligned}
\tag{72}
$$

Without loss of generality, we identify $\gamma_1(\mu, 0) = -\mu \in \mathbb{R}$, and we set $\gamma_2(0,0) = 1$. Nontrivial patterns in $\alpha = 0$ only exist for $\mu > 0$. We therefore restrict to $\mu > 0$. We may now scale $A = |\mu|^{1/2}\tilde{A}$, $B = |\mu|\tilde{B}$, $r = |\mu|^{-1/2}\tilde{r}$, and $\alpha = \tilde{\alpha}|\mu|^{1/2}$. The scaled equation is

$$
\begin{aligned}
\tilde{A}_{\tilde{r}} &= \tilde{B} - \frac{\tilde{\alpha}}{2}\tilde{A} + |\mu|^{1/2}\tilde{\mathcal{R}}_1(\tilde{A}, \tilde{B}, |\mu|^{1/2}\tilde{\alpha}; \mu) \\
\tilde{B}_{\tilde{r}} &= -\frac{\tilde{\alpha}}{2}\tilde{B} - \tilde{A} + \tilde{A}|\tilde{A}|^2 + |\mu|^{1/2}\tilde{\mathcal{R}}_2(\tilde{A}, \tilde{B}, |\mu|^{1/2}\tilde{\alpha}; \mu) \\
\tilde{\alpha}_{\tilde{r}} &= -\tilde{\alpha}^2
\end{aligned}
\tag{73}
$$

with $\tilde{\mathcal{R}}_j$ bounded in its arguments, uniformly in $\tilde{\alpha} \leq \mu^{-1/2}\alpha_0$.

We do not discuss the case of a weakly subcritical Turing instability any further.

4.2. Heteroclinics in the universal equations. We show the existence of transverse heteroclinics in the scaled and truncated equations (70), (71), and (73), with μ set to zero. Since the complete section refers to the scaled equations only, we drop tildes.

Recall that the scaled equations on $\tilde{\mathcal{W}}^{\mathrm{c}}_{+}$ are valid for large spatial times, as $\mu \to 0$. In $\alpha = 0$, we find equilibria, case (O), and periodic orbits, case (T). They possess a stable (O) or center-stable (T) manifold $\tilde{W}^{\mathrm{s/cs}}_{+}$ inside $\tilde{\mathcal{W}}^{\mathrm{c}}_{+}$, which contains the solutions converging to those patterns as $r \to \infty$. We transport this manifold backwards in the radius r inside the manifold $\tilde{\mathcal{W}}^{\mathrm{c}}_{+}$. For $\mu \to 0$, the scaled radial time $\tilde{r}$ spent on $\tilde{\mathcal{W}}^{\mathrm{c}}_{+}$, $\alpha \leq \alpha_0$, converges to infinity. We therefore consider the scaled equation formally at $\mu = 0$, for all $\tilde{r} \geq 0$. The scaled equations possess well-defined asymptotics for $\tilde{r} \to 0$. In case of a homogeneous instability (O), the dynamics in a neighborhood of $r \to 0$ are very similar to the dynamics in the full equation. The situation in the Turing instability (T) is slightly different.

Consider the fold first, (70), with $\mu = 0$:

$$A' = B, \quad B' = -\alpha B + 1 - A^2, \quad \alpha' = -\alpha^2.$$

Note that this is the equation for radially symmetric patterns of the *scalar* elliptic equation

$$\triangle A - 1 + A^2 = 0.$$

Let $W^{\mathrm{cu}}_{-}(r)$ denote the set of initial values at time r, which lead to bounded solutions as $r \to 0$. From Proposition 3.5 we know that $W^{\mathrm{cu}}_{-}(r)$ is a smooth, one-dimensional manifold for each $r > 0$.

At $\alpha = 0$, we have two equilibria from the fold, $A = \pm 1$, $B = 0$. Inside $\alpha = 0$, only $A = -1$ is hyperbolic (and actually stable for the reaction-diffusion system (2)). Denote by $W^{\mathrm{s}}_{+}(r;-1)$ the initial values which lead to solutions converging to $(A,B) = (-1,0)$ for $r \to \infty$. From Proposition 3.6 and hyperbolicity of $(A,B) = (-1,0)$ in the asymptotic equation for $r = \infty$, we know that $W^{\mathrm{s}}_{+}(r;-1)$ is a one-dimensional manifold for each r as well.

PROPOSITION 3.15. [**KP97**] *There exists a unique nonconstant bounded solution $q_*(r)$ to (70) with $\mu = 0$, which converges to $A_* = -1$ for $r \to \infty$. The solution is smooth down to $r = 0$ and transverse:*

$$\mathcal{T}_{q_*(r)} W^{\mathrm{s}}_{+}(r;-1) \oplus \mathcal{T}_{q_*(r)} W^{\mathrm{cu}}_{-}(r) = \mathbb{R}^2.$$

Proof. We refer to [**KP97**] for a comprehensive discussion of the general type of problem. There are actually several ways to prove the proposition. One way is to exploit the variational structure of the equation. We outline another proof using a shooting argument. The existence proof via a shooting argument goes back to [**BLP81**]; see also [**KP97**, Thm. 24.1]. First observe, that the Hamiltonian for the $\alpha = 0$-system

$$J(A,B) = \frac{1}{2}B^2 - A + \frac{1}{3}A^3$$

is a Lyapunov functional for the nonautonomous system:

$$\frac{\mathrm{d}}{\mathrm{d}r} J(A(r),B(r)) = -\frac{1}{r}B^2 \leq 0.$$

The nonautonomous term $\frac{1}{r}B$ acts like a damping to the pendulum equation in $\alpha = 0$. Note, that $J(-1,0) = 2/3$ and the level set $J = 2/3$ consists of the stable and unstable manifolds of $(-1,0)$; see also Chapter 2, Figure 3. In particular, the interior of the bounded subset of $\mathbb{R}^2$ inside the homoclinic is forward invariant under the evolution to the nonautonomous equation. We consider the trajectories $(A(r), B(r))$ in $\tilde{W}_-^{\mathrm{cu}}$, that is $A(0) = A_0$, $B(0) = 0$. We trace the first zero of B for $r > 0$ when A_0 is increased. For $A_0 > 0$ small, we stay inside the region bounded by the homoclinic and we have $B(r_0(A_0)) = 0$ for some finite r_0. One also easily checks that $r_0' < 0$. On the other hand, for $A_0 >> 1$ large, there is no zero of B. The initial value A_0^* of the heteroclinic q_* is found as the supremum over initial values A_0, for which there still exists r_0 with $B(r_0) = 0$. Transversality is proved using monotonicity of the zero $r_0' < 0$; see also [**KH81b**]. ∎

Consider the cusp, next. The equation at $\mu_2 = 0$

$$\begin{aligned} \tilde{A}_{\tilde{r}} &= \tilde{B} \\ \tilde{B}_{\tilde{r}} &= -\tilde{\alpha}\tilde{B} + \tilde{\mu_1} - \tilde{A} + \tilde{A}^3 \end{aligned}$$

is as above the equation for radially symmetric solutions of a scalar elliptic equation

$$\triangle A - \mu_1 + A - A^3 = 0.$$

The asymptotic equation (71) at $\alpha = 0$, with $\mu_2 = 0$ becomes

$$A' = B, \quad B' = -\alpha B + \mu_1 - A + A^3, \quad \alpha' = -\alpha^2.$$

It possesses three equilibria $A_- < A_0 < A_+$, $B = 0$, within the bistability region, bounded by $|\mu_1| < \frac{2}{3\sqrt{3}} = \mu_*$; see Proposition 2.24. The outer equilibria $(A_\pm, 0)$ are saddles in the asymptotic equation (71) with $\alpha = 0, \mu_2 = 0$. They possess stable manifolds which we denote by $W_+^{\mathrm{s}}(r; A_\pm)$.

PROPOSITION 3.16. *For $-\mu_* < \mu_1 < 0$, there exists a unique, smooth, nonconstant bounded solution $q_*^+(r)$ to (71) with $\mu_2 = 0$, which for $r \to \infty$ converges to $(A_+, 0)$. For $0 < \mu_1 < \mu_*$, there exists a unique nonconstant bounded solution $q_*^-(r)$ to (71) with $\mu = 0$, which for $r \to \infty$ converges to $(A_-, 0)$. The solution is smooth down to $r = 0$ and transverse: $\mathcal{T}_{q_*^\pm(r)} W_+^{\mathrm{s}}(r; A_\pm) \oplus \mathcal{T}_{q_*^\pm(r)} W_-^{\mathrm{cu}}(r) = \mathbb{R}^2$.*

Proof. [**KP97**] The proof is similar to the proof of Proposition 3.15.

We give a different proof for μ_1 close to zero, which emphasizes the aspect of coexistence of stable equilibria. The idea of the proof will be further exploited in Chapter 4 in order to find target patterns. In the region of μ_1 small, we may interprete the solution as a phase boundary between the two stable states $A_\pm$. We find the heteroclinics from a heteroclinic bifurcation, which is depicted in Figure 3. For $\mu_1 = 0$, we have a heteroclinic cycle in the system at $\alpha = 0$, connecting A_+ to A_- and vice versa. In addition, we have transverse heteroclinics connecting $r = 0$ to $\alpha = 0$ along the constants $A_\pm = \pm 1$. By transversality and the λ-Lemma [**KH95**, Thm. 6.2.8, Prop. 6.2.23], $\tilde{W}_-^{\mathrm{cu}}$ is close to the center-unstable manifolds of $(A_\pm, 0)$. The center-stable manifold of the equilibria $(A_\pm, 0)$ intersect transversely the center-unstable manifolds of $(A_\mp, 0)$ along the asymptotic heteroclinic cycle in the extended phase space. Tangent vectors in the center direction are transported with the linearization, where $\alpha' = 0$ and the derivative with respect to α is the same as for the damped pendulum, where transversality is known; see Chapter 2, Chapter 2, Section 3.4. Varying μ_1, this transverse intersection persists as an intersection for $\alpha > 0$ or $\alpha < 0$ depending on the sign of μ_1. Since $\tilde{W}_-^{\mathrm{cu}}$ is close to the center-unstable manifolds of the equilibria, we obtain heteroclinic intersections between $\tilde{W}_-^{\mathrm{cu}}$ and $\tilde{W}_+^{\mathrm{cs}}(A_\pm)$. The connections appear for different signs of μ_1, since the equation is invariant under the reflection $(A, B; \mu_1) \mapsto (-A, -B; -\mu_1)$. ∎

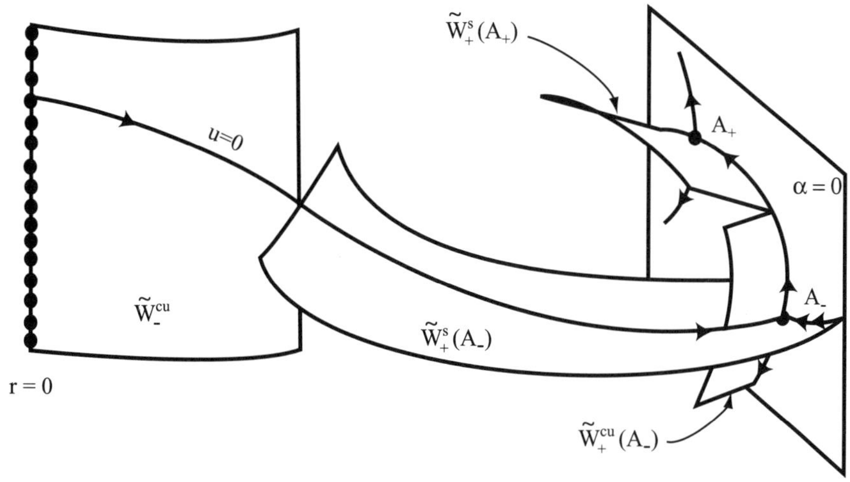

FIGURE 3. *The coexistence boundary in the cusp bifurcation is created from two heteroclinics, representing the trivial first state of the system in radial dynamics, and the one-dimensional coexistence interface.*

Consider the generic, supercritical Turing instability, which has lead us to the reduced, scaled equation (73). We rewrite (73) at $\mu = 0$, setting $\hat{B} = B - \frac{\alpha}{2}A$, as

$$\begin{aligned} A_r &= \hat{B} \\ \hat{B}_r &= -\alpha\hat{B} + \frac{1}{4}\alpha^2 A - A + A|A|^2 \\ \alpha_r &= -\alpha^2 \end{aligned} \tag{74}$$

Note that this formally is the equation for charge-1/2 defects to the Ginzburg-Landau equation

$$\triangle A + A - A|A|^2 = 0, \tag{75}$$

that is, solutions of the form $A(r)\mathrm{e}^{\mathrm{i}\ell\varphi}$ with topological charge $\ell = 1/2$ — which, of course, would be discontinuous as a solution to (75).

In the $\alpha = 0$-fiber, we have three real equilibria with $A = -1, 0,$ and 1. The equilibrium with $A_+ = 1$ is hyperbolic within the real subspace and we denote by $\tilde{W}^{cs}_+(A_+)$ its center-stable manifold in the extended phase space $\mathbb{C}^2 \times \mathbb{R}^+$.

We also need to consider $\tilde{r}$ small due to the scaling in μ for r small. We set $\tau = \log r$ and $\hat{B} = r^{-1}B^0$ and obtain

$$\begin{aligned} A_\tau &= B^0 \\ B^0_\tau &= \frac{1}{4}A + r^2(-A + A|A|^2) \\ r_\tau &= r. \end{aligned} \tag{76}$$

The equilibrium $A = B^0 = 0$, $r = 0$, possesses a two-dimensional center-unstable manifold which we denote by W^{cu}_-.

PROPOSITION 3.17. *There exists a unique bounded, real solution $q_*(r)$ of (74, 76), which converges to $(A, B) = (1, 0)$ for $r \to \infty$ in (74) and is contained in W_-^{cu}, with $q_*(r) = \mathrm{O}(\sqrt{r})$. The heteroclinic is transverse in the full complex phase space of (74), that is, W_-^{cu} and $W_+^{\mathrm{cs}}(A_+)$ intersect transversely along q_*.*

Proof. Is identical to the one in [**KH81b**, Lem. 3.1]. There, the equation $A'' + A/r - k^2/r^2 A = -A + A|A|^2$ was considered with k integer. The case $k = 1/2$, here, does not alter the proofs, given there. ∎

Similarly, transverse heteroclinics can be found for the subcritical and weakly subcritical equation.

4.3. Persistence. We show that the heteroclinics found in Section 4.2, in the scaled equation within $\tilde{\mathcal{W}}_+^{\mathrm{c}}$, yield branches of solutions in the full reaction diffusion system. The main step consists in a matching procedure at the boundary $\alpha = \alpha_0$ of $\tilde{\mathcal{W}}_+^{\mathrm{c}}$. Main tool is again transversality, exploiting in addition the particular scalings.

We start analyzing the case of the fold.

THEOREM 3.18. [Fold] *Assume a linearly generically unfolded homogeneous instability (O), Definition 2.10. Assume that the bifurcation in the kinetics is a fold with two equilibria for $\mu > 0$; see Section 4.1. Then the solution of the universal equation found in Proposition 3.15 yields a unique branch of solutions to the full reaction-diffusion system (44), asymptotic for $r \to \infty$ to a spatially homogeneous equilibrium of amplitude $\sqrt{|\mu|}$. The solutions are smooth in r down to $r = 0$.*

In terms of the original reaction-diffusion system, we have found a branch of nontrivial, radially symmetric solutions $U(r; \mu)$, of amplitude $\mathrm{O}(\sqrt{|\mu|})$, emanating from the origin $U \equiv 0$, $\mu = 0$. The nontrivial solutions coexist with the two trivial, spatially homogeneous equilibria which coalesce in the fold bifurcation. The solution $U(r; \mu)$ converges for $r \to \infty$ to the spatially homogeneous equilibrium which is stable for the reaction-diffusion system.

Proof. We want to find intersections of the center-unstable manifold $\tilde{\mathcal{W}}_-^{\mathrm{cu}}$ with the stable manifold $\tilde{W}_+^{\mathrm{s}} \subset \tilde{\mathcal{W}}_+^{\mathrm{c}}$. We therefore try to match the two manifolds at $r = r_* = 1/\alpha_0$. We first compute expansions of the center-unstable manifold $W_-^{\mathrm{cu}}(r_*) = \mathcal{W}_-^{\mathrm{cu}}(r_*) \cap \tilde{\mathcal{W}}_+^{\mathrm{c}}$, projected on the tangent space to $\tilde{\mathcal{W}}_+^{\mathrm{c}}$ at $A = B = 0$, $\alpha = \alpha_0$, the center eigenspace E_+^{c}. We claim that $W_-^{\mathrm{cu}}(r_*) = \{(a, \psi_-^{\mathrm{u}}(a)); a \in (-\delta, \delta)\}$ with some finite δ small and $\psi_-^{\mathrm{u}}(a) = \mathrm{O}(a^2 + |\mu|)$, in the coordinates of Lemma 3.9. Indeed, the tangent space of $\mathcal{W}_-^{\mathrm{cu}}(r_*)$ in $\underline{u} = 0$ is transported by the linearized equation. Within the subspace E_+^{c}, which is invariant for the linearized equation, the solutions that are bounded at $r = 0$ are of the form $(a, 0)$, which proves the claim. We next compute the expansion of the stable manifold $\tilde{W}_+^{\mathrm{s}}(r_*)$ of $\tilde{A} = \sqrt{|\mu|}$, inside $\tilde{\mathcal{W}}_+^{\mathrm{c}}(r_*)$, projected on E_+^{c}; see Proposition 3.15 for the definition of the manifold $\tilde{W}_+^{\mathrm{s}}$. In the scaled coordinates, $r = r*$ corresponds to $\tilde{r}_* = |\mu|^{1/4} r_*$, which is close to zero.

We argue next that the stable manifold $\tilde{W}_+^{\mathrm{s}}(r_*)$ is close to the unstable subspace $E_-^{\mathrm{cu}}(r_*)$. We therefore first consider the truncated, reduced equation. We transport in backwards radial time the stable manifold $\tilde{W}_+^{\mathrm{s}}(r_*)$, which is transverse to the center-unstable manifold of the origin, $W_-^{\mathrm{cu}}(r)$, for all r by transversality of the heteroclinic. Evoking the λ-lemma, any such transverse manifold converges to the stable subspace at $r = 0$, with exponential rate $1 - \epsilon$ in the rescaled radial time τ.

Note that the origin is not hyperbolic, but the λ-lemma still holds [**KH95**, Thm. 6.2.8, Prop. 6.2.23]. In particular, at scaled radial time $\tilde{r}_* = |\mu|^{1/4} r_*$, the stable manifold $\tilde{W}^{\mathrm{s}}_+$ at $\tilde{r}_* = |\mu|^{1/4} r_*$ is $|\mu|^{1/4-\epsilon}$-close to the unstable subspace at $r = 0$. In the scaled coordinates, it is given through

$$(77)\quad \tilde{W}^{\mathrm{s}}_+(r_*) = \{(\tilde{q}(0) + \psi^{\mathrm{s}}_+(b), b); b \in (-\delta, \delta)\}, \quad \text{with } \psi^{\mathrm{s}}_+(b) = \mathrm{O}(|\mu|^{1/4-\epsilon} + |b|^2)$$

for any fixed $\epsilon > 0$ small. We estimate the perturbation coming from the remainder terms, next. Integrating the perturbation $\mathrm{O}(\sqrt{|\mu|})$ over a time-interval $\mathrm{O}(|\mu|^{1/4})$, gives a perturbation of at most $\mathrm{O}(|\mu|^{1/4})$. This does not alter the expansion in (77).

In the original, un-scaled coordinates, the expansion (77) transforms into

$$(78)\qquad W^{\mathrm{s}}_+(r_*) = \left\{ \left(|\mu|^{1/2} \left(\tilde{q}(0) + \psi^{\mathrm{s}}_+(b;\mu) \right), |\mu|^{3/4} b \right); b \in (-\delta, \delta) \right\}.$$

We look for intersections between $W^{\mathrm{s}}_+(r_*)$ and $W^{\mathrm{cu}}_-(r_*)$, which are solutions to

$$a = |\mu|^{1/2}(\tilde{q}(0) + \psi^{\mathrm{s}}_+(b)), \quad \psi^{\mathrm{u}}_-(a) = |\mu|^{3/4} b.$$

Scaling $a = |\mu|^{1/2}\tilde{a}$ gives

$$\tilde{a} = \tilde{q}(0) + \psi^{\mathrm{s}}_+(b), \quad b = |\mu|^{-3/4}\psi^{\mathrm{u}}_-(\tilde{a}|\mu|^{1/2} a, \mu).$$

By the expansion for ψ^{u}_-, the second equation becomes $b = \mathrm{O}(|\mu|^{1/4})$, and we can solve for $(\tilde{a}, b)(\mu)$ with the implicit function theorem. ∎

The cusp bifurcation allows for an analogous treatment.

THEOREM 3.19. [Cusp] *Assume a linearly generically unfolded homogeneous instability (O), Definition 2.10. Assume that the bifurcation in the kinetics is a cusp with two stable equilibria inside the cuspoidal region; see Section 4.1. Then the solution of the universal equation found in Proposition 3.16 inside the cuspoidal region yields a unique branch of solutions to the full reaction-diffusion system (44), asymptotic for $r \to \infty$ to a spatially homogeneous equilibrium of amplitude $\sqrt{|\mu_2|}$. The solutions are smooth in r down to $r = 0$.*

In terms of the original reaction-diffusion system, we have found nontrivial radially symmetric solutions $U(r;\mu)$, of amplitude $\mathrm{O}(\sqrt{|\mu_2|})$, close to the origin $U \equiv 0$, $\mu = 0$. The nontrivial solutions exist in the cuspoidal region of coexistence of stable equilibria, outside the curve in parameter space, where the two stable equilibria coexist in the one-dimensional problem, Proposition 3.16. The nontrivial solution $U(r;\mu)$ converges to the spatially homogeneous equilibrium for $r \to \infty$, which is stable for the reaction-diffusion system, but which in the one-dimensional problem is invaded by the other stable equilibrium; see Remark 2.26. The theorem does not guarantee the existence of radially symmetric patterns inside the entire coexistence domain, although we strongly suspect that this is what actually happens.

Proof. The proof is similar to the proof of Theorem 3.18. We indicate the necessary changes. The expansion for $W^{\mathrm{cu}}_-(r_*) = \{(a, \psi^{\mathrm{u}}_-(a)); a \in (-\delta, \delta)\}$ is $\psi^{\mathrm{u}}_-(a) = \mathrm{O}(|a|^3 + |\mu_2 a| + |\mu_2|^2)$, since for $\mu_1 = 0$, the trivial equilibrium persists, and for $\mu = 0$, quadratic terms vanish. The expansion for $W^{\mathrm{s}}_+(r_*)$ in scaled coordinates is the same as in the case of a fold , (77). The perturbation term $\mathcal{R}$ in (71) again only gives a small contribution. In the original scaling, we obtain $W^{\mathrm{s}}_+(r_*) = \{(|\mu_2|^{1/2}(\tilde{q}(0) + \psi^{\mathrm{s}}_+(b), \mu), |\mu_2 | b\}; b \in (-\delta, \delta)\}$. We scale as before $a = |\mu_2|^{1/2}\tilde{a}$ and solve the equation for the intersection

$$\tilde{a} = \tilde{q}(0) + \psi^{\mathrm{s}}_+(b), \quad b = |\mu_2|^{-1}\psi^{\mathrm{u}}_-(|\mu_2|^{1/2}\tilde{a}).$$

In particular, $b = \mathrm{O}(|\mu_2|^{1/2})$ and we may solve the equation by the implicit function theorem. ∎

In case of a Turing instability, a similar persistence result holds.

THEOREM 3.20. [Turing] *Assume a linearly generically unfolded Turing instability (T), Definition 2.10. Assume that the cubic coefficient in the reduced equation is positive, that is, stable Turing patterns of amplitude* $\mathrm{O}(\sqrt{\mu})$ *exist in the region* $\mu > 0$*, where the trivial state is unstable. Then the solution of the universal scaled equation, found in Proposition 3.17, yields a unique branch of solutions to the full equation for any fixed asymptotic wavenumber* k *close enough to* k_**. The solutions are smooth in* r *down to* $r = 0$*.*

The solutions found in this theorem correspond to radially symmetric focus patterns of the original reaction-diffusion system; see Chapter 1, Figure 5. For large values of the radius, they resemble concentric circles, spaced at the uniform distances $2\pi/k$. Up to the curvature terms $\mathrm{O}(1/r)$, they consist of essentially one-dimensional Turing patterns. Close to the center, the amplitude of the Turing pattern decays. The construction of the pattern is as a transverse heteroclinic orbit between the origin, $\mathcal{W}^{\mathrm{cu}}_{-}$, and the stable manifold of a single asymptotic Turing pattern. By robustness, for any fixed parameter value μ, there exists a family of such focus patterns, parameterized by the asymptotic wavenumber k. We do not know if focus patterns exist for any small amplitude asymptotic Turing pattern. The results presented here do only give information for patterns which are close to the most critical wavenumber k_*. Probably, Eckhaus stability of the asymptotic Turing pattern is a necessary condition for existence.

Proof. We look for intersections of the manifold $\tilde{\mathcal{W}}^{\mathrm{cu}}_{-}$ that consists of solutions of the full system (44), which are bounded for $r \to 0$, with the stable manifold $\tilde{W}^{\mathrm{s}}_{+} \subset \tilde{\mathcal{W}}^{\mathrm{c}}_{+}$ of the Turing pattern $(\tilde{A}, \tilde{B}) = (1, 0)$ from Proposition 3.17, inside the asymptotic center manifold $\tilde{\mathcal{W}}^{\mathrm{c}}_{+}$. We find these intersections for fixed radial time $r_* = 1/\alpha_*$, sufficiently large, independent of μ. We therefore compute Taylor expansions of these manifolds.

To start with, we compute expansions for the tangent space of $\tilde{\mathcal{W}}^{\mathrm{cu}}_{-}$ in $\underline{u} = 0$. We therefore consider the full reaction-diffusion system (44). Going back to the proof of Lemma 3.10, we already have seen, that the tangent space E^{c}_{+} of $\tilde{\mathcal{W}}^{\mathrm{c}}_{+}$ in $\underline{u} = 0$ is independent of r, and the linearized reaction-diffusion equation (46) can be written in Jordan normal-form

$$u_1'' = -\frac{1}{r}u_1' - k_*^2 u_1 + u_2, \qquad u_2'' = -\frac{1}{r}u_2' - k_*^2 u_2. \tag{79}$$

Without loss of generality, we set $k_* = 1$ in the sequel, rescaling the radius r if necessary.

Solutions bounded as $r \to 0$ are given explicitly in terms of Bessel functions. A way of finding expansions for these solutions is, to start with bounded or mildly growing solutions to the equation for (u_1, u_2) omitting the terms with $1/r$. We find $u_1(r) = \cos r$ and $u_1(r) = r \sin r$. We then substitute x_1 for r and consider u_1 as a solution for the two-dimensional problem, independent of x_2, where $x = (x_1, x_2)$. These are smooth solutions to the two-dimensional problem. Rotating these functions around the origin, we find new solutions. The average of these solutions over all possible rotations gives a radially symmetric, bounded solutions to (79). Expliciting the average integrals, we find explicit bases for the component of the tangent space to $\tilde{\mathcal{W}}^{\mathrm{cu}}_-$ in E^{c}_+ in terms of Bessel functions:

$$\underline{u}^1(r) = (J_0(r), 0, J_0'(r), 0)^T, \quad \underline{u}^2(r) = (rJ_1(r), J_0(r), (rJ_1)'(r), 2J_0'(r))^T$$

where the J_k denote the Bessel functions of the first kind.

Exploiting the asymptotic expansions

$$J_0(r) = \frac{1}{\sqrt{r}}\left(\cos(r - \frac{\pi}{4}) + \mathrm{O}(\frac{1}{r})\right), \quad J_1(r) = \frac{1}{\sqrt{r}}\left(\cos(r - \frac{3\pi}{4}) + \mathrm{O}(\frac{1}{r})\right),$$

we find

$$\begin{aligned}
\underline{u}^1(r) &= \frac{1}{\sqrt{r}}\left(\cos(r - \frac{\pi}{4}), 0, \cos(r + \frac{\pi}{4}), 0\right)^T + \mathrm{O}(r^{-3/2}) \\
\underline{u}^2(r) &= \sqrt{r}\left(\cos(r - \frac{3\pi}{4}) + \mathrm{O}(\frac{1}{r}), \frac{1}{r}\cos(r - \frac{\pi}{4}),\right. \\
&\qquad \left.\cos(r - \frac{\pi}{4}) + \mathrm{O}(\frac{1}{r}), \frac{2}{r}\cos(r + \frac{\pi}{4})\right)^T + \mathrm{O}(r^{-3/2}).
\end{aligned}$$

In complex (A, B)-coordinates, this gives, after renormalization of the basis,

$$(A, B) = \mathrm{e}^{\mathrm{i}\varphi}(s_1 + \mathrm{i}s_2 + \mathrm{O}(\frac{1}{r}), \frac{1}{r}\mathrm{i}s_2) + \mathrm{O}(r^{-2}),$$

where $\varphi = r - \frac{\pi}{4}$. The nonlinear manifold can therefore be written in the form

$$\tilde{W}^{\mathrm{cu}}_- = \left\{(A, B)^T = \mathrm{e}^{\mathrm{i}\varphi}\left(s_1 + \mathrm{i}s_2 + \mathrm{O}(\frac{1}{r}), \frac{1}{r}\mathrm{i}s_2\right)^T + \mathrm{O}(r^{-2} + s_1^2 + s_2^2)\right\}.$$

Consider next the matching subspace $\tilde{W}^{\mathrm{s}}_+$. We scale the reduced equation (72) for $\mu > 0$ as in Section 4.1, $A = \sqrt{\mu}a$, $B = \sqrt{\mu}b$, $\alpha = \sqrt{\mu}\beta$, $\partial_r = \sqrt{\mu}\partial_\rho$. For convenience, we use the slightly different notation avoiding tildes. We find

$$\begin{aligned}
a_\rho &= b - \frac{\beta}{2}a + \mathrm{O}(\sqrt{\mu}) \\
b_\rho &= -a - \frac{\beta}{2}b + a|a|^2 + \mathrm{O}(\sqrt{\mu}) \\
\beta_\rho &= -\beta^2.
\end{aligned}$$

Error terms are uniformly bounded in $\beta \le \beta_0\mu^{-1/2}$, together with their derivatives. We reparameterize the radius $\mathrm{e}^\tau = \rho$ and factor out the diagonal linear part, replacing $\hat{a} = \sqrt{\rho}a$, $\hat{b} = \sqrt{\rho}b$. We find, to leading order

$$\begin{aligned}
\hat{a}_\tau &= \rho\hat{b} \\
\hat{b}_\tau &= -\rho\hat{a} + \hat{a}^3 \\
\rho_\tau &= \rho,
\end{aligned}$$

up to remainder terms $\mathrm{O}(\sqrt{\mu})$. Note that $\rho = 0$ is invariant with flow

$$\hat{a}(\tau) = \hat{a}(0), \quad \hat{b}(\tau) = \hat{b}(0) + \tau(\hat{a}(0))^3.$$

LEMMA 3.21. *The invariant plane $\rho = 0$ is smoothly fibered by a strong unstable fibration: for any $\delta > 0$ and $\rho_0 > 0$, there exists a smooth coordinate change*

$(\hat{a},\hat{b}) \mapsto (\tilde{a},\tilde{b})$, *depending on* ρ, $0 \le \rho \le \rho_0$, $\mathrm{O}(\rho_0^{1-\delta})$*-close to the identity, such that in the new coordinates we have*

$$\tilde{a}_\tau = 0, \quad \tilde{b}_\tau = \tilde{a}^3, \quad \rho_\tau = \rho.$$

Proof. *[of Lemma 3.21]* We renormalize $\hat{a} = a_0 + a_\mathrm{n}$, $\hat{b} = b_0 + a_0^3\tau + b_\mathrm{n}$ and obtain

$$\begin{aligned}\dot{a}_\mathrm{n} &= \rho\hat{b} = \mathrm{e}^\tau(b_0 + a_0^3\tau + b_\mathrm{n})\\ \dot{b}_\mathrm{n} &= -\rho\hat{a} + \hat{a}^3 - a_0^3 = -\mathrm{e}^\tau(a_0 + a_\mathrm{n}) + (a_0 + a_\mathrm{n})^3 - a_0^3\end{aligned}$$

and $\dot{} = \frac{\mathrm{d}}{\mathrm{d}t}$ denotes derivative with respect to time $\tau = \log\rho$. Trajectories which decay in backwards time are found as fixed points of the integral formulation

$$\begin{aligned}a_\mathrm{n}(\tau) &= \int_{-\infty}^{\tau}\left(\mathrm{e}^\sigma(b_0 + \sigma a_0^3 + b_\mathrm{n}(\sigma))\right)\mathrm{d}\sigma\\ b_\mathrm{n}(\tau) &= \int_{-\infty}^{\tau}\left(-\mathrm{e}^\sigma(a_0 + a_\mathrm{n}(\sigma)) + (a_0 + a_\mathrm{n}(\sigma))^3 - a_0^3\right)\mathrm{d}\sigma.\end{aligned}$$

The right side defines a map on functions $a_\mathrm{n}, b_\mathrm{n} \in C^0((-\infty,\tau_0],\mathbb{R})$. Equipping the continuous functions with exponential weights

$$\|(a,b)\|_\delta = \sup_{\tau\le\tau_0}\mathrm{e}^{(1-\delta)\tau}(|a(\tau)| + |b(\tau)|),$$

we readily show that the right side defines a contraction mapping in a sufficiently small neighborhood of $a \equiv b \equiv 0$. Dependence on a_0 and b_0 is smooth, since the nonlinearity is differentiable on the function space. ∎

We continue the proof of Theorem 3.20. Recall, that we consider W_+^s in the neighborhood of the basic solution $q_*(\rho)$ from Proposition 3.17 which decays for $\rho \to 0$ with expansion $q_*(\rho) = q_*^0\rho^{1/2} + \mathrm{O}(\rho^{3/2})$. We know from Proposition 3.17 that this particular solution is a transverse heteroclinic. In particular, we do not have solutions to the real, linearized equation with a bounded a-component

$$a_{\rho\rho} + \frac{1}{\rho}a_\rho - \frac{1}{4\rho^2}a + a - 3q_*^2(\rho)a = 0.$$

We expand W_+^s in the coordinates $\tilde{a},\tilde{b}$ from Lemma 3.21. Since $a \to 0$ in the original coordinates, $\hat{a},\tilde{a} \to 0$. By Lemma 3.21, we therefore have $\tilde{b} \equiv const =: q_*^0$. We therefore can parameterize part of the stable manifold by

$$W_+^\mathrm{s}(\rho) = \left\{(0,q_*^0) + (\psi_1^\mathrm{s}(\lambda),\psi_2^\mathrm{s}(\lambda)),\ \lambda\in\mathbb{C})\right\}.$$

Since the linearized equation does not possess bounded solutions, we know that have $(\psi_1^\mathrm{s})'(0)$ is invertible. Indeed, otherwise the tangent space to W_+^s would be given by $(0,*)$ and solutions in this tangent space would remain bounded for the linearized equation. We can therefore rewrite

$$W_+^\mathrm{s}(\rho) = \{(\lambda,\psi^\mathrm{s}(\lambda)),\ \lambda\in\mathbb{C}\}.$$

We transport this manifold with the flow from Lemma 3.21 in backwards time $\rho \to 0$ and derive the expansion

$$W_+^\mathrm{s}(\rho) = \left\{(\lambda, \mathrm{O}(|\lambda| + |\lambda|^3|\log\rho|),\ \lambda\in\mathbb{C}\right\}.$$

Back in (a,b)-coordinates, rescaling $\lambda = \rho^{1/2}\tilde{\lambda}$, this gives $W_+^\mathrm{s}(\rho) = \{(a,b) = (\tilde{\lambda},\mathrm{O}(|\tilde{\lambda}|)),\ \lambda\in\mathbb{C}\}$. Rescaling with μ to the normal form $(\tilde{A},\tilde{B})$-coordinates, we find the expansion $W_+^\mathrm{s}(\rho) = \{(\tilde{A},\tilde{B}) = (\lambda|\mu|^{1/2},0) + \mathrm{O}(|\mu|^{3/4}),\ \lambda\in\mathbb{C}\}$. In the last step, we go back the normal form transformation, which however is readily seen to preserve the subspace $(*, 0)$ to first order in α and we find in the original (A, B)-coordinates

$$W_+^s(\rho) = \left\{(A, B) = \left(\lambda|\mu|^{1/2}, \mathrm{O}(\alpha^2|\mu|^{1/2} + |\mu|^{3/4})\right), \ \lambda \in \mathbb{C}\right\}.$$

We have to match the expressions we have obtained for W_+^s and W_-^{cu} so far. Intersections solve the system of equations

$$\begin{aligned} \lambda|\mu|^{1/2} &= (s_1 + \mathrm{i}s_2)\mathrm{e}^{\mathrm{i}\varphi} + \mathrm{O}(\alpha_*(|s_1| + |s_2|) + s_1^2 + s_2^2) \\ \alpha_* \mathrm{i}s_2 \mathrm{e}^{\mathrm{i}\varphi} &= \mathrm{O}\left(\alpha_*^2(|s_1| + |s_2|) + \alpha_*^2|\mu|^{1/2} + |\mu|^{3/4}\right), \end{aligned}$$

where $\alpha_* = 1/r_*$, small, is the matching distance to the origin and $\varphi = 1/\alpha_* - \pi/4$. In order to solve this system of equations, we fix $s_1 = 0$ and vary $\alpha_* > 0$, small, instead. Rescaling $s_2 = |\mu|^{1/2}\sigma$ gives

$$\begin{aligned} \lambda &= \mathrm{i}\sigma\mathrm{e}^{\mathrm{i}\varphi} + \mathrm{O}(\alpha_*|\sigma| + |\mu|^{1/2}\sigma^2) \\ \alpha_*\mathrm{i}\sigma\mathrm{e}^{\mathrm{i}\varphi} &= \mathrm{O}(\alpha_*^2 + |\mu|^{1/4}). \end{aligned}$$

We solve the first equation for λ and plug the result into the second equation:

$$\mathrm{i}\sigma\mathrm{e}^{\mathrm{i}/\alpha_*} = \mathrm{O}\left(\alpha_*|\sigma| + |\mu|^{1/4}\right).$$

Since the linearization of the left hand side of this equation is invertible and the right hand side is small, we find locally unique solution branches in μ. The perturbations $\mathcal{R}_j$ in (73) are again of lower order along the solution. This proves Theorem 3.20. ∎

REMARK 3.22. *The analysis* and *the result differ from the formal arguments in* [**BS78, PZM85, CH93**]. *We do not use phase-diffusion equations, which loose their validity in the core region; see also* [**Schn95**]. *Instead we treat the full reduced problem, which is equivalent to a Swift-Hohenberg equation. The result proves existence of patterns with* bounded *amplitude, in spite of the (necessarily) diverging wave vector B/A. A second difference comes from the considerations of nonadiabatic effects. The phase-diffusion equation includes an averaging over the finite-wavelength Turing patterns in the spirit of the normal form transformation in Theorem 2.17. Near the center of the focus pattern, this averaging breaks down. The effect is seen already at the level of a formal dimension counting argument. We prove existence of focus patterns for an open range of asymptotic wave numbers k: in the nonautonomous radial dynamics, the stable manifold of a Turing pattern at $r = \infty$ is two-dimensional and intersects the two-dimensional matching manifold W_-^{cu} transversely. The dimension counting argument is in fact much more generally valid, not only for patterns of small amplitude near homogeneous equilibria. Amplitude equations, like the universal equation (76) we derived, predict a wave number selection by the focus pattern* [**CH93**].

CHAPTER 4

Time-periodic radially symmetric patterns

This chapter is devoted to the existence of time-periodic, radially symmetric solutions of the reaction-diffusion system (2) close to an oscillatory instability (H), (TH). Combining the ideas from Chapter 2, Section 4, where we used ill-posed, one-dimensional spatial dynamics on time-periodic functions to describe small bounded solutions, with the reduction procedure from Chapter 3 for radial dynamics, we prove center manifold reductions, analogous to Theorems 3.3 and 3.7, for radial dynamics on time-periodic functions; see Theorems 4.1 and 4.8. The starting point is a formulation of the problem in radial dynamics in Section 1. In Section 2, the main reduction results are proved. Roughly following the lines of Chapter 3, we derive far-field equations on $\tilde{\mathcal{W}}^{\mathrm{c}}_{+}$ in case (H), Section 3, and discuss heteroclinic orbits for the scaled equations in Section 4. We conclude the section with the main existence results, Theorems 4.12 and 4.13 on existence of target patterns in the full reaction-diffusion system. Again, the final step is a matching procedure between manifolds $\tilde{W}^{\mathrm{cu}}_{-}$ from the core region and $\tilde{W}^{\mathrm{s}}_{+}$ from the asymptotic plane waves. The plane waves in the far-field of the target-pattern solutions in the theorems possess group velocity directed towards the center in Theorem 4.12, and group velocity directed away from the center in Theorem 4.13. The latter are found in a degenerate bifurcation, where we recover a cubic-quintic complex Ginzburg-Landau equation. We refer to the discussion, Chapter 5, Section 3, for an analysis of target patterns in the generic, cubic case.

We do not attempt to describe solutions of the reduced equations in case (TH), although focus patterns in the spirit of Theorem 3.20, asymptotic to standing waves, might be amenable to the type of analysis that we present here.

We generally stay with the notation introduced in Chapter 3, Section 2.1.

1. Radial dynamics on time-periodic functions

We study oscillatory instabilities of

$$U_t = D\triangle U + F(U;\mu),$$

where $F(0;\mu) = 0$ allows for a trivial homogeneous equilibrium for μ close to zero; see Chapter 2, Section 2. The linearized equation

$$V_t = D\triangle V + \partial_U F(0;0)V,$$

is assumed to possess the unique, up to reflection in x and complex conjugation, nondecaying solution

$$V(t,x) = \mathrm{e}^{\mathrm{i}\omega_* t}\mathrm{e}^{\pm \mathrm{i}k_* x}U_0,$$

with $\omega_* \neq 0$, see Definition 2.3 for the definition of oscillatory instabilities of type (H) and (TH). Following the strategy of Chapter 2, Section 4, we rescale time

$t = \omega^{-1}\tilde{t}$ with parameter ω close to ω_*, and look for 2π-periodic solutions, which are radially symmetric:

$$\omega U_t = D\left(U_{rr} + \frac{n-1}{r}U_r\right) + F(U;\mu), \quad U(0,r) = U(2\pi,r),\ U_r(0,r) = U_r(2\pi,r),$$

where $r = |x|$ is the radius in polar coordinates. We rewrite the system as a differential equation in spatial time r

$$(80) \qquad \begin{aligned} u_r &= w \\ w_r &= -\frac{n-1}{r}w - D^{-1}\left(-\omega\partial_t u + F(u;\mu)\right) \end{aligned}$$

with linearization

$$(81) \qquad \begin{aligned} v_r &= w \\ w_r &= -\frac{n-1}{r}w - D^{-1}\left(-\omega\partial_t v + \partial_U F(0;0)v\right). \end{aligned}$$

Here $\underline{u}(r,\cdot) = (u,w)(r,\cdot)$ and $\underline{v}(r,\cdot) = (v,w)(r,\cdot)$ are 2π-periodic functions of time t. We briefly write (81) as $\underline{u}_r = \mathcal{A}(r;\omega)\underline{u}$. The nonlinear equation can be written shortly as $\underline{u}_r = \mathcal{A}(r;\omega)\underline{u} + \mathcal{F}(\underline{u};\mu)$ with $\mathcal{F}((u,v);\mu) = (0, -D^{-1}(F(u;\mu) - \partial_U F(0;0)u))^T = \mathrm{O}(|u|^2 + |\mu|)$.

The linearized equation decouples into infinitely many ordinary differential equations using the Fourier decomposition $(v,w) = \sum_{\ell\in\mathbb{Z}}(v^\ell, w^\ell)\mathrm{e}^{\mathrm{i}\ell t}$. We obtain

$$(82) \qquad \begin{aligned} v_r^\ell &= w^\ell \\ w_r^\ell &= -\frac{n-1}{r}w^\ell - D^{-1}\left(-\omega\mathrm{i}\ell v^\ell + \partial_U F(0;0)v^\ell\right). \end{aligned}$$

For $r \to \infty$, we formally obtain an asymptotic equation which is precisely the same as in the one-dimensional case, Chapter 2, Section 4:

$$(83) \qquad \begin{aligned} v_r^\ell &= w^\ell \\ w_r^\ell &= -D^{-1}\left(-\omega\mathrm{i}\ell v^\ell + \partial_U F(0;0)v^\ell\right). \end{aligned}$$

We write (83) in the compact form $\underline{v}_r^\ell = \mathcal{A}^\ell(\infty;\omega)\underline{v}^\ell$. Denote by $E_+^{\ell,\mathrm{c}}$ the center eigenspace of $\mathcal{A}^\ell(\infty;\omega_*)$, and let $E_+^{\mathrm{c}} := \sum_\ell E_+^{\ell,\mathrm{c}}\mathrm{e}^{\mathrm{i}\ell t} \leq Y = H^{1/2}(S^1,\mathbb{R}^N) \times L^2(S^1,\mathbb{R}^N)$ denote the center eigenspace of $\mathcal{A}(\infty;\omega_*) : Y^1 \subset Y \to Y$; see also Chapter 2, Section 4 for the definitions of the function spaces Y, Y^1. Recall from Chapter 2, Section 4 that for $|\ell|$ sufficiently large, $E_+^{\ell,\mathrm{c}}$ is trivial.

Similarly, we construct infinite-dimensional stable, E_+^{s}, and unstable, E_+^{u}, subspaces of the operator $\mathcal{A}(\infty;\omega_*) : Y^1 \subset Y \to Y$.

These spaces can be continued to spaces $E_+^{\mathrm{s}}(r_*)$, $E_+^{\mathrm{c}}(r_*)$, $E_+^{\mathrm{u}}(r_*)$, which represent initial conditions to (83) giving rise to strongly continuous solution operators $\Phi_+^{\mathrm{s,c,u}}(r,r_*) : Y \to Y$, mapping $E_+^j(r_*)$ onto $E_+^j(r)$, on $1 \leq r_* \leq r$, on $r_*, r \geq 1$, and on $1 \leq r \leq r_*$, respectively. Moreover, we have the uniform estimates

$$(84) \qquad |\Phi_+^{\mathrm{s,u}}(r,r_*)|_{Y\to Y} \leq C\mathrm{e}^{-\eta|r-r_*|}, \quad |\Phi_+^{\mathrm{c}}(r,r_*)|_{Y\to Y} \leq C\mathrm{e}^{\delta|r-r_*|}$$

for any $\delta > 0$ with some constants $C(\delta)$ and $\eta > 0$. These solution operators can be constructed for each ℓ separately, see Lemma 3.1 and [**SS99**] (a different approach was taken in [**PSS97**]).

In the asymptotic equation, setting $r=\infty$, reversibility again acts as a reflection $R:(u,w)\mapsto(u,-w)$; see Chapter 3, Section 2.1. Note that $RE^{\mathrm{s}}_{+}=E^{\mathrm{u}}_{+}$ and that $\dim E^{\mathrm{c}}_{+}=d^{\mathrm{c}}<\infty$ is finite and even, by reversibility.

Similar to Chapter 3, we consider the autonomous systems

$$\begin{aligned} u' &= w \\ w' &= -(n-1)\alpha w - D^{-1}\left(-\omega\partial_t u + F(u;\mu)\right) \\ \alpha' &= -\alpha^2, \end{aligned} \tag{85}$$

with $'=\frac{\mathrm{d}}{\mathrm{d}r}$, for $\alpha\le\alpha_*$ finite, describing the limit $r\to\infty$, and

$$\begin{aligned} \dot u &= rw \\ \dot w &= -(n-1)w - D^{-1}\left(-\omega r\partial_t u + rF(u;\mu)\right) \\ \dot r &= r \end{aligned} \tag{86}$$

with $\dot{}=\frac{\mathrm{d}}{\mathrm{d}\tau}$, $\tau=\log r$, describing the limit $r\to 0$. Note however, that in the formal limiting system, (86) with $r=0$, the differential operator ∂_t disappears.

2. Center manifolds

2.1. The nonautonomous center manifold $\tilde{\mathcal{W}}^{\mathrm{c}}$. Following the ideas from Chapter 3, Section 2.1, we construct a nonautonomous center manifold, which contains the set of bounded solutions. The following result refers to the definition of a nonautonomous center manifold, Definition 3.2, with the concept of a *solution* as in Chapter 2, Section 4.2: $\underline{u}\in C^0(J,Y)$ is a solution on an interval J, if $\underline{u}\in C^1(\operatorname{int}J,Y)\cap C^0(\operatorname{int}J,Y^1)$ and $\underline{u}_r=\mathcal{A}(r;\omega)\underline{u}+\mathcal{F}(\underline{u};\mu)$ for $r\in\operatorname{int}J$. Recall that temporal time-shifts act on Y via γ_θ, $\theta\in\mathbb{R}/2\pi\mathbb{Z}$; see Chapter 2, Section 4.

THEOREM 4.1. *For each $0<m<\infty$ and μ close to zero, there exists a C^m-center manifold $\tilde{\mathcal{W}}^{\mathrm{c}}$ for equations (85), (86), of dimension $\frac{1}{2}\dim E^{\mathrm{c}}_{+}+1$. The dependence on the parameter μ is C^m. The manifold is invariant and the local flow on the manifold is equivariant under the action of the temporal time-shift symmetry γ_θ.*

The proof follows the proof of Theorem 3.3. We construct manifolds $\tilde{W}^{\mathrm{cu}}_{-}$ and $\tilde{\mathcal{W}}^{\mathrm{cs}}_{+}$ for a modified equation, which are backward and forward invariant, respectively. We then define $\tilde{\mathcal{W}}^{\mathrm{c}}$ as the intersection $\tilde{\mathcal{W}}^{\mathrm{cu}}_{-}\cap\tilde{\mathcal{W}}^{\mathrm{cs}}_{+}$, restricted to a small neighborhood of $\underline{u}=0$.

We start modifying F as in Chapter 3, Section 2.1 with the smooth cut-off function χ, for some small enough δ',

$$F_{\mathrm{mod}}(u;\mu):=\chi\left(\frac{|u|^2+|\mu|^2}{\delta'}\right)\left(F(u;\mu)-\partial_U F(0;0)u\right),$$

with $\chi(s)=1$ for $s<1$ and $\chi(s)=0$ for $s\ge 2$, $\chi'<0$ on $(1,2)$. The associated modified nonlinearity $\mathcal{F}_{\mathrm{mod}}((u,w);\mu)=(0,-D^{-1}F_{\mathrm{mod}}(u;\mu))$ is globally Lipschitz continuous on Y with $\operatorname{Lip}\mathcal{F}_{\mathrm{mod}}\to 0$ for $\delta'\to 0$.

We construct the center-stable manifold $\tilde{\mathcal{W}}^{\mathrm{cs}}_{+}$, first. Consider the modified version of equation (85), describing solutions near $r=\infty$,

$$\begin{aligned} u' &= w \\ w' &= -(n-1)\alpha w - D^{-1}\left(-\omega\partial_t u + F_{\mathrm{mod}}(u;\mu)\right) \\ \alpha' &= -\alpha^2. \end{aligned} \tag{87}$$

and $(u, w, \alpha) \in Y \times \mathbb{R}$.

DEFINITION 4.2. We say a manifold $\tilde{\mathcal{M}} \subset Y \times \mathbb{R}$ is a C^m global center-stable manifold of (87), if $\tilde{\mathcal{M}} \subset Y \times \{|\alpha| \leq \alpha_0\}$ for some $\alpha_0 > 0$, and if there is a global strongly continuous semi-flow $\tilde{\Phi}^{\rm cs}(r)$, $r \geq 0$ on $\tilde{\mathcal{M}}$, such that $\tilde{\mathcal{M}}$ is forward invariant under $\tilde{\Phi}^{\rm cs}$, trajectories of $\tilde{\Phi}^{\rm cs}$ are solutions of (87), and $\tilde{\mathcal{M}}$ contains all solutions of (87), which are defined for all $r \geq 1/\alpha_0$ and bounded as $r \to \infty$.

PROPOSITION 4.3. *For any μ close to zero and ω close to ω_*, and for any $0 < m < \infty$, there is $\alpha_0 > 0$ such that (87) possesses a C^m global center-stable manifold $\tilde{\mathcal{W}}^{\rm cs}_+$ with global semi-flow $\tilde{\Phi}^{\rm cs}_+(r)$, $r \geq 0$, see Definition 4.2. Moreover, $\tilde{\mathcal{W}}^{\rm cs}_+$ is tangent to $(E^{\rm c}_+ \oplus E^{\rm s}_+) \times \mathbb{R}$ at $\underline{u} = 0$, $\alpha = 0$, and $\mu = 0$, $\omega = \omega_*$ in the extended phase space $Y \times \mathbb{R}$. Also, $(\underline{u} = 0, \alpha = 1/r_*) \in \tilde{\mathcal{W}}^{\rm cs}_+$, and the tangent space at this point is $(E^{\rm s}(r_*) \oplus E^{\rm c}(r_*)) \times \mathbb{R}$ for $\mu = 0$. Also, $\tilde{\mathcal{W}}^{\rm cs}_+$ is δ'-close to $Y \times \{|\alpha| \leq \alpha_0\}$ in the C^m-topology.*

The manifold $\tilde{\mathcal{W}}^{\rm cs}_+$ is invariant and the semi-flow $\tilde{\Phi}^{\rm cs}_+(r)$ is equivariant under the time-shift symmetry γ_θ.

Proof. The proof is similar to the proof for center-stable manifolds in ordinary differential equations in [**Van89**]. Modifications for infinite-dimensional problems as considered here are given in [**VI91**] in a very general context. However, the discussion in [**VI91**] restricts to center manifolds. We give a short outline of the proof for center-stable manifolds, here.

We modify the equation for α as in the proof of Proposition 3.4 to $\alpha' = -\chi(\alpha/\delta')\alpha^2$ where χ is the smooth cut-off function from above, and obtain an equation which is a δ'-small perturbation of the linearized asymptotic equation

$$\underline{u}' = \mathcal{A}(\infty;\omega)\underline{u}, \quad \alpha' = 0. \tag{88}$$

This linear equation allows for a spectral decomposition in a center-stable subspace $\tilde{E}^{\rm cs} := (E^{\rm s} \oplus E^{\rm c}_+) \times \mathbb{R}$, and an unstable subspace $\tilde{E}^{\rm u} := E^{\rm u}_+ \times \{0\}$. We write $\tilde{P}^{\rm cs}$ for the continuous projection on $\tilde{E}^{\rm cs}$ along $\tilde{E}^{\rm u}$. Solutions of (88) are given (explicitly) from the decomposition into Fourier subspaces (83). The initial value problem to (88) for initial data $\tilde{\underline{u}}(0) = \tilde{\underline{u}}_0$ can be solved for $(\underline{u}_0, \alpha) \in \tilde{E}^{\rm cs}$ for $r > 0$ and we denote the linear solution operator by $(\underline{u}(r), \alpha(r)) = \tilde{\Phi}^{\rm cs}_+(r)\tilde{\underline{u}}_0$. For $(\underline{u}_0, \alpha) \in \tilde{E}^{\rm u}$, the initial value problem can be solved for $r < 0$ and we denote the linear solution operator by $(\underline{u}(r), \alpha(r)) = \tilde{\Phi}^{\rm u}_+(r)\tilde{\underline{u}}_0$. There are $\eta^{\rm u} > 0$, $C > 0$, and for any $\delta > 0$, there is $C' > 0$ such that

$$\begin{aligned} |\tilde{\Phi}^{\rm cs}_+(r)| &\leq C'{\rm e}^{\delta r}, \ r \geq 0; \\ |\tilde{\Phi}^{\rm u}_+(r)| &\leq C{\rm e}^{\eta^{\rm u} r}, \ r \leq 0 \end{aligned}$$

as operators in $\tilde{Y} = Y \times \mathbb{R}$. The solutions of (87) with weak exponential growth as $r \to \infty$ are now found as solutions of a variation-of-constant formula. Write $\tilde{\underline{u}} = (\underline{u}, \alpha) \in Y \times R$, and

$$\tilde{G}(u, w, \alpha) = \left(0, -D^{-1}F_{\rm mod}(u;\mu), -\chi(\alpha/\delta')\alpha^2\right)^T.$$

Then consider the fixed point equation

$$\tilde{\underline{u}}(r) = \tilde{\Phi}^{\rm cs}_+(r)\tilde{P}^{\rm cs}\tilde{\underline{u}}_0 + \int_0^r \tilde{\Phi}^{\rm cs}_+(r-s)\tilde{P}^{\rm cs}\tilde{G}(\tilde{\underline{u}}(s)){\rm d}s + \int_\infty^r \tilde{\Phi}^{\rm u}_+(r-s)(1-\tilde{P}^{\rm cs})\tilde{G}(\tilde{\underline{u}}(s)){\rm d}s$$

in the space $C^0_\eta(\mathbb{R}_+,\tilde{Y})$ with norm

$$|\underline{\tilde{u}}(\cdot)| := \sup_{r\geq 0} \mathrm{e}^{-\eta r}\|\underline{\tilde{u}}(r)\|_{\tilde{Y}}.$$

With m being the degree of differentiability in the proposition, we choose $\eta^{\mathrm{u}} > m\eta > m\delta$ and δ' small. Then the right side of the variation-of constant formula, viewed as an operator on $C^0_\eta(\mathbb{R}_+,\tilde{Y})$, defines a contraction. The unique fixed point $\underline{\tilde{u}}(r)$, depending on the parameter $\underline{\tilde{u}}_0 = \tilde{P}^{\mathrm{cs}}\underline{\tilde{u}}(0)$, gives $\tilde{\mathcal{W}}^{\mathrm{cs}}_+$ as the set of points $\underline{\tilde{u}}(0)$, with $|\alpha| \leq \alpha_0 = \delta'$. The nonlinear semi-flow Φ^{cs} on $\tilde{\mathcal{W}}^{\mathrm{cs}}_+$ is defined as associating the solution $\underline{\tilde{u}}(r)$ to the initial value $\underline{\tilde{u}}_0$. Differentiability and invariance properties follow as in the case for ODEs; see for example [**Van89**]. Differentiability with respect to ω can be seen by rescaling the equation as in Chapter 2, Section 4 and incorporating ω in the parameter μ. Tangent spaces are transported by the linearized equation at $\underline{u} = 0$, which possesses unique r-dependent center-stable subspaces $E^{\mathrm{cs}}_+(r)$, as mentioned in Section 1. ∎

REMARK 4.4. *Viewing (88) as a small perturbation of its nonautonomous linearization $\underline{u}_r = \mathcal{A}(r;\omega)\underline{u}$, we may actually construct $\tilde{\mathcal{W}}^{\mathrm{cs}}_+$ on any fixed but finite domain $|\alpha| \leq \alpha_0 < \infty$, α_0 not necessarily small. We therefore replace $\tilde{\Phi}^{\mathrm{cs}}$ and $\tilde{\Phi}^{\mathrm{u}}$ by their nonautonomous counter parts, the evolution operators to (88). The center-stable manifold is then again constructed as a fixed point of the above variation-of-constant formula. This is a nontrivial extension of Proposition 4.3, since we cannot globalize locally invariant manifolds by means of transporting them with a flow, as the initial value problem is ill-posed and solutions to the initial-value problem typically do not exist.*

As a next step, we construct $\mathcal{W}^{\mathrm{cu}}_-$. We start analyzing the linearized equation

$$\begin{aligned} v' &= w \\ w' &= -\frac{n-1}{r}w - D^{-1}\left(-\omega\partial_t v + \partial_U F_{\mathrm{mod}}(0;0)v\right), \end{aligned} \tag{89}$$

or, in spatial time $\tau = \log r$,

$$\begin{aligned} \dot{v} &= \mathrm{e}^{\tau} w \\ \dot{w} &= -(n-1)w - \mathrm{e}^{\tau} D^{-1}\left(-\omega\partial_t v + \partial_U F_{\mathrm{mod}}(0;0)v\right). \end{aligned} \tag{90}$$

LEMMA 4.5. *Equation (90) possesses an exponential dichotomy with rates $\eta^{\mathrm{s}} = n-1$ and $\eta^{\mathrm{u}} = 0$ on $\tau \in (-\infty,\tau_0]$, for any $\tau_0 < \infty$. More precisely, there are constants C^{s} and C^{u} and strongly continuous linear evolution operators $\Phi^{\mathrm{s}}_-(\tau,\sigma)$ for $\tau_0 \geq \tau \geq \sigma$ and $\Phi^{\mathrm{u}}_-(\tau,\sigma)$ for $\tau_0 \geq \sigma \geq \tau$, such that $\Phi^{\mathrm{s}}_-(\cdot,\sigma)\underline{v}$ and $\Phi^{\mathrm{u}}_-(\cdot,\sigma)\underline{v}$ are solutions of (90), for any $\underline{v} \in Y$. Moreover, $\Phi^{\mathrm{s}}_-(\sigma,\sigma)\underline{v} + \Phi^{\mathrm{u}}_-(\sigma,\sigma)\underline{v} = \underline{v}$ and*

$$\begin{aligned} \|\Phi^{\mathrm{s}}_-(\tau,\sigma)\|_{\mathcal{L}(Y,Y)} &\leq C^{\mathrm{s}}\mathrm{e}^{-(n-1)|\tau-\sigma|},\ \tau_0 \geq \tau \geq \sigma, \\ \|\Phi^{\mathrm{u}}_-(\tau,\sigma)\|_{\mathcal{L}(Y,Y)} &\leq C^{\mathrm{u}},\ \tau_0 \leq \tau \leq \sigma. \end{aligned}$$

Proof. Since the linearized equation decouples into an infinite product of ordinary differential equations for the Fourier modes, it suffices to show the claim for each Fourier mode $\mathrm{e}^{\mathrm{i}\ell t}$, with ℓ-uniform constants $C^{\mathrm{s/u}}_\ell$.

Therefore, consider

$$\begin{aligned} \dot{v}^\ell &= \mathrm{e}^{\tau} w^\ell \\ \dot{w}^\ell &= -(n-1)w^\ell - \mathrm{e}^{\tau} D^{-1}\left(-\omega \mathrm{i}\ell v^\ell + \partial_U F_{\mathrm{mod}}(0;0)v^\ell\right), \end{aligned} \tag{91}$$

on $\mathbb{C}^{2N}$ with norm $|(v^\ell, w^\ell)|_\ell^2 := (|\ell|+1)|v^\ell|^2+|w^\ell|^2$. For each ℓ fixed, (91) possesses an exponential dichotomy as claimed, since the asymptotic equations $\dot{v}^\ell = 0$, $\dot{w}^\ell = -(n-1)w^\ell$ possess exponential dichotomies and the coefficients of the perturbations decay exponentially in τ for $\tau \to -\infty$. We have to consider $|\ell| \geq \ell_0$ large and show uniformity of constants with respect to the norm $|\cdot|_\ell$. We may assume $\ell > 0$, the other case being complex conjugate. We set $\tilde{v}^\ell = \sqrt{\ell}v^\ell$ such that $|(v^\ell, w^\ell)|_\ell \simeq |(\tilde{v}^\ell, w^\ell)|$ define uniformly equivalent norms. We obtain

$$\begin{aligned}
\text{(92)} \qquad \frac{\mathrm{d}}{\mathrm{d}\tau}\tilde{v}^\ell &= \sqrt{\ell}\,\mathrm{e}^\tau w^\ell \\
\frac{\mathrm{d}}{\mathrm{d}\tau}w^\ell &= -(n-1)w^\ell - \mathrm{e}^\tau D^{-1}\left(-\omega\mathrm{i}\sqrt{\ell}\tilde{v}^\ell + \sqrt{\ell}^{-1}\partial_U F_{\mathrm{mod}}(0;0)\tilde{v}^\ell\right).
\end{aligned}$$

The term involving F is a small perturbation for ℓ large, with uniform exponential decay in τ, and by robustness of dichotomies [**Cop78, PSS97**], it is sufficient to consider

$$\begin{aligned}
\text{(93)} \qquad \frac{\mathrm{d}}{\mathrm{d}\tau}\tilde{v}^\ell &= \sqrt{\ell}\mathrm{e}^\tau w^\ell \\
\frac{\mathrm{d}}{\mathrm{d}\tau}w^\ell &= -(n-1)w^\ell - \mathrm{e}^\tau D^{-1}(-\omega\mathrm{i}\sqrt{\ell}\tilde{v}^\ell).
\end{aligned}$$

We write $(\tilde{v}^\ell, w^\ell) = ((\tilde{v}_1^\ell, \ldots, \tilde{v}_N^\ell)^T, (w_1^\ell, \ldots, w_j^\ell)^T) \in \mathbb{C}^N \times \mathbb{C}^N$, expliciting the N species of the reaction-diffusion system. Now, in (93), the equations for the individual species $(\tilde{v}_j^\ell, w_j^\ell)$ decouple and we can consider each component separately. We scale $\hat{v}_j^\ell = (d_j/\omega)^{-1/2}\mathrm{i}^{1/2}\tilde{v}_j^\ell$, where d_j is the j'th entry in the diffusion matrix D, and translate $\sigma = \tau + \log\sqrt{\ell} - \log\sqrt{d_j/\omega}$ to obtain

$$\begin{aligned}
\text{(94)} \qquad \frac{\mathrm{d}}{\mathrm{d}\sigma}\hat{v}_j^\ell &= \mathrm{i}^{1/2}\mathrm{e}^\sigma w_j^\ell \\
\frac{\mathrm{d}}{\mathrm{d}\sigma}w_j^\ell &= -(n-1)w_j^\ell + \mathrm{i}^{1/2}\mathrm{e}^\sigma\hat{v}_j^\ell.
\end{aligned}$$

We have to show that there exist exponential dichotomies on $\sigma \in (-\infty, \log\sqrt{\ell})$. We show that the ℓ-independent equation (94) indeed possesses an exponential dichotomy on the entire real line $\mathbb{R}$, which, together with the previous considerations, proves the lemma.

First, observe that (94) possesses an exponential dichotomy on $\mathbb{R}^-$, since we recover a linear equation with eigenvalues 0 and $-(n-1)$ for $\sigma \to -\infty$. Also, note that (94) does not possess nontrivial, bounded solutions. Indeed, any solution of (94) is a solution of $v'' + (n-1)v'/r - \mathrm{i}v = 0$. Solutions, bounded for $r \to 0$, are Bessel functions of the first kind with complex argument, which are known to grow exponentially as $r \to \infty$.

It therefore suffices to show that (94) possesses an exponential dichotomy on $\mathbb{R}^+$. We first consider (94) in the new time-variable $s = \mathrm{e}^\sigma$:

$$\begin{aligned}
\text{(95)} \qquad \frac{\mathrm{d}}{\mathrm{d}s}\hat{v}_j^\ell &= \mathrm{i}^{1/2}w_j^\ell \\
\frac{\mathrm{d}}{\mathrm{d}s}w_j^\ell &= -\frac{n-1}{s}w_j^\ell + \mathrm{i}^{1/2}\hat{v}_j^\ell.
\end{aligned}$$

Since the asymptotic equation at $s=\infty$, $\frac{\mathrm{d}^2}{\mathrm{d}s^2}\hat{v}_j^\ell = i\hat{v}_j^\ell$ possesses an exponential dichotomy, we have an exponential dichotomy on $s \geq 1$ for (95):

$$\begin{aligned}|\tilde{\Phi}^{\mathrm{s}}(s_1,s_2)| &\leq C^{\mathrm{s}}\mathrm{e}^{-\frac{1}{2}|s_1-s_2|}, \quad \text{for } s_1 \geq s_2 \geq 1\\ |\tilde{\Phi}^{\mathrm{u}}(s_1,s_2)| &\leq C^{\mathrm{u}}\mathrm{e}^{-\frac{1}{2}|s_1-s_2|}, \quad \text{for } s_2 \geq s_1 \geq 1.\end{aligned}$$

Translating back to σ-time gives evolution operators

$$\Phi^{\mathrm{s/u}}(\sigma_1,\sigma_2) := \tilde{\Phi}^{\mathrm{s/u}}(\mathrm{e}^{\sigma_1},\mathrm{e}^{\sigma_2}),$$

with exponential decay estimate

$$|\Phi^{\mathrm{s/u}}(\sigma_1,\sigma_2)| \leq C^{\mathrm{u}}\mathrm{e}^{-\frac{1}{2}|\mathrm{e}^{\sigma_1}-\mathrm{e}^{\sigma_2}|} \leq C^{\mathrm{u}}e^{-c|\sigma_1-\sigma_2|},$$

where $c = \mathrm{e}^{-\frac{1}{2}\min(\sigma_1,\sigma_2)}$.

Summarizing, we have found exponential dichotomies for (91), uniformly in ℓ. Exponential decay of the nonautonomous part $-\mathrm{e}^{\tau}D^{-1}\partial_U F_{\mathrm{mod}}(0;0)v^\ell$, also uniform in ℓ, implies that the exponential rates coincide with the rates $-(n-1)$ and 0, predicted by the equation in $\tau=-\infty$. This proves existence of dichotomies with exponential estimates as claimed. ∎

We write $E_-^{\mathrm{cu}}(r) := \mathrm{Rg}\,(\Phi_-^{\mathrm{u}}(\log r,\log r))$ for the center-unstable subspace. By construction, these subspaces are unique. They can be continued, using the flow in the Fourier subspaces to $\tau\in\mathbb{R}$. The union of the center-unstable subspaces in the extended phase space $\bigcup_{r>0}(E_-^{\mathrm{cu}}(r),r)$ is denoted by $\tilde{E}_-^{\mathrm{cu}}$.

DEFINITION 4.6. We say a manifold $\tilde{\mathcal{M}} \subset Y\times\mathbb{R}$ is a C^m global center-unstable manifold of (86), if $\tilde{\mathcal{M}} \subset Y\times\{r\leq r_0\}$ for some $r_0>0$, and if there is a global backward semi-flow $\tilde{\Phi}_-^{\mathrm{cu}}(r)$, $r\leq 0$ on $\tilde{\mathcal{M}}$, such that $\tilde{\mathcal{M}}$ is backward invariant under $\tilde{\Phi}_-^{\mathrm{cu}}$, trajectories of $\tilde{\Phi}_-^{\mathrm{cu}}$ are solutions of (86), and $\tilde{\mathcal{M}}$ contains all solutions of (86), which are defined for all $r\leq r_0$ and bounded as $r\to 0$.

PROPOSITION 4.7. *For any $0<m<\infty$, $\mu\in\mathbb{R}^p$, $\omega>0$, and $r_0\in\mathbb{R}^+$, there is a C^m global center-unstable manifold $\tilde{\mathcal{W}}_-^{\mathrm{cu}}$ to the equilibrium $\underline{u}=0$, $r=0$ in equation (86), see Definition 4.6, which depends C^m on μ and ω. The manifold is tangent to $E_-^{\mathrm{cu}}(r)\times\mathbb{R}$ at $\underline{u}=0$, $0<r\leq r_0$, and $\mu=0$, $\omega=\omega_*$, and it is δ'-close to $\tilde{E}_-^{\mathrm{cu}}$ in the C^m-topology. Trajectories of the backwards semi-flow $\tilde{\Phi}_-^{\mathrm{cu}}$ on $\tilde{\mathcal{W}}_-^{\mathrm{cu}}$ yield solutions of (86).*

The manifold $\tilde{\mathcal{W}}_-^{\mathrm{cu}}$ is invariant and the semi-flow $\tilde{\Phi}_-^{\mathrm{cu}}(r)$ is equivariant under the time-shift symmetry γ_θ.

In fact, all backward-trajectories of $\tilde{\Phi}_-^{\mathrm{cu}}$ are bounded and convergent, and all bounded solutions of (86) on $\tau\in(-\infty,\log r_0]$, $r\leq r_0$, are trajectories of $\tilde{\Phi}_-^{\mathrm{cu}}$.

Proof. We use the information on the linearized problem from Lemma 4.5 to set up a fixed point argument, similar to the proof of Proposition 4.3. However, we construct the manifold in the phase space Y, not in the extended phase space $Y\times\mathbb{R}$. We set

$$G(u,w,\tau) = \left(0, -\mathrm{e}^{\tau}D^{-1}(F_{\mathrm{mod}}(u;\mu) - \partial_U F_{\mathrm{mod}}(0;0)u)\right)^T.$$

Any bounded solution $\underline{u}(\tau)$ of (86) is a solution of the variation-of-constant formula

$$\underline{u}(\tau) = \Phi_-^{\mathrm{u}}(\tau,\tau_0)\underline{u}_0 + \int_{\tau_0}^{\tau}\Phi_-^{\mathrm{u}}(\tau,\sigma)G(\underline{u}(\sigma),\sigma)\mathrm{d}\sigma + \int_{-\infty}^{\tau}\Phi_-^{\mathrm{s}}(\tau,\sigma)G(\underline{u}(\sigma),\sigma)\mathrm{d}\sigma.$$

Recall from Lemma 4.5 that the norm of the operators $\Phi^{\mathrm{s}}_{-}(\tau,\sigma)$ decays exponentially in $\tau-\sigma>0$, and that the norm of the operators $\Phi^{\mathrm{u}}_{-}(\tau,\sigma)$ is uniformly bounded in $\tau-\sigma<0$. We also have exponential decay of the nonlinearity G:

$$|G(\underline{u}^1(\sigma),\sigma)-G(\underline{u}^2(\sigma),\sigma)|_Y \le \mathrm{O}(\delta')\mathrm{e}^{\sigma}|\underline{u}^1-\underline{u}^2|_{BC^0((-\infty,\tau_0],Y)}.$$

With these ingredients, it is straight forward to show that the right side defines a contraction on $BC^0((-\infty,\tau_0],Y)$. The unique fixed point $\underline{u}(\tau;\underline{u}_0)$ with $\underline{u}_0\in E^{\mathrm{cu}}_{-}(\mathrm{e}^{\tau_0})$ defines the center-unstable manifold with center-unstable flow. Convergence for $r\to 0$ of individual trajectories is also an easy consequence of the variation-of-constant formula, together with the above estimates. ∎

Proof. *[of Theorem 4.1]* We define $\tilde{\mathcal{W}}^{\mathrm{c}}_{\mathrm{glob}}=\tilde{\mathcal{W}}^{\mathrm{cu}}_{-}\cap\tilde{\mathcal{W}}^{\mathrm{cs}}_{+}$. We have to show transversality of the intersection, which reduces to showing that

(i) $E^{\mathrm{cu}}_{-}(r)\cap E^{\mathrm{cs}}_{+}(r)=:E^{\mathrm{c}}(r)$ is of dimension $\dim E^{\mathrm{c}}_{+}(r)$;
(ii) $E^{\mathrm{cu}}_{-}(r)+E^{\mathrm{cs}}_{+}(r)=Y$.

Step (i) follows as in the proof of Theorem 3.3, considering the linearized equation for each Fourier mode separately. Bounded solutions only occur within the Fourier subspace $\ell=1$.

We focus on (ii), next. We have to show that $E^{\mathrm{cu}}_{-,\ell}(r)$ and $E^{\mathrm{cs}}_{+,\ell}(r)$ are transverse to each other, uniformly in ℓ for large ℓ. However, for ℓ large, these subspaces are close to $\mathcal{E}^{\mathrm{cu}}_{-,\ell}(r)$ and $\mathcal{E}^{\mathrm{cs}}_{+,\ell}(r)$, the corresponding subspaces for the slightly perturbed equation (93)

$$\begin{aligned}\frac{\mathrm{d}}{\mathrm{d}r}\tilde{v}^{\ell} &= \sqrt{\ell}w^{\ell}\\ \frac{\mathrm{d}}{\mathrm{d}r}w^{\ell} &= -\frac{n-1}{r}w^{\ell}-D^{-1}\left(-\omega\mathrm{i}\sqrt{\ell}\tilde{v}^{\ell}\right),\end{aligned}$$

see Lemma 4.5 and its proof. Now, this equation was already shown to possess an exponential dichotomy on $\mathbb{R}$ in Lemma 4.5. There, the equation was considered in logarithmic spatial time $\tau=\log r$, which, however, does not alter the definition of subspaces. This proves transversality, uniformly in ℓ and, with the previous discussion, Theorem 4.1. ∎

2.2. The asymptotic center manifold $\tilde{\mathcal{W}}^{\mathrm{c}}_{+}$. For large radii, we can construct an asymptotic center manifold $\tilde{\mathcal{W}}^{\mathrm{c}}_{+}$ as in Chapter 3, Section 2.2.

THEOREM 4.8. *For each $0<m<\infty$, μ close to zero, and ω close to ω_*, there exists a C^m-center manifold $\tilde{\mathcal{W}}^{\mathrm{c}}_{+}$ for (85) near $\underline{u}=0$, $\alpha=0$, with local flow $\tilde{\Phi}^{\mathrm{c}}_{+}$. The center manifold $\tilde{\mathcal{W}}^{\mathrm{c}}_{+}$ is tangent to $E^{\mathrm{c}}_{+}\times\mathbb{R}$ in the extended phase space $(\underline{u},\alpha)\in Y\times\mathbb{R}$ at the point $\underline{u}=0$, $\alpha=0$, and $\mu=0$, $\omega=\omega_*$.*

Moreover, $\tilde{\mathcal{W}}^{\mathrm{c}}_{+}$ contains all small bounded solutions: there are $\alpha_0>0$ and $\delta'>0$, such that, if $|\underline{u}(r)|\le\delta'$ for all $r>0$ is a solution, then $(\underline{u}(r),1/r)\in\tilde{\mathcal{W}}^{\mathrm{c}}_{+}$ for all $r\ge 1/\alpha_0$.

The center manifold is invariant and the flow $\tilde{\Phi}^{\mathrm{c}}_{+}$ is equivariant under the action γ_θ of temporal time-shifts.

The construction is the same as in the proof of Theorem 3.7. We construct the center manifold within $\mathcal{W}^{\mathrm{cs}}_{+}$, where we can use the semi-flow $\tilde{\Phi}^{\mathrm{cs}}_{+}$ to define the graph transform.

3. The reduced vector field for a Hopf instability

We derive the general normal form of the vector field on $\mathcal{W}^{c}_{+}$ in case of a Hopf instability (H).

First, consider the linear part. We choose coordinates (A, B) as in Lemma 2.30. There are no constant terms for all α, since $(A, B) \equiv 0$ is a solution of the full nonautonomous equation. For the linearized equation, E^{c}_{+} is invariant. This can be seen as in the proof of Lemma 3.9. Within E^{c}_{+}, α-dependence is as in the stationary homogeneous instability, and we arrive at the expansion

$$(96)\qquad \begin{aligned} A_r &= B + \mathrm{O}((|A| + |B|)^3) \\ B_r &= \gamma_1(\omega, \mu)A - (n-1)\alpha B + \mathrm{O}\left((|A| + |B|)^3\right) \\ \alpha_r &= -\alpha^2 \end{aligned}$$

with γ_1 as in Proposition 2.32. Note that quadratic terms vanish due to the phase invariance caused by the temporal time-shift symmetry.

Further exploiting the time-shift symmetry and normal form transformations as described in Proposition 3.11, we are lead to the expansion

$$(97)\qquad \begin{aligned} A_r &= B + \mathrm{i}AP_1\left(|A|^2, \mathrm{i}(A\overline{B} - \overline{A}B), \alpha; \omega, \mu\right) + \mathcal{R}_1(A, B, \alpha; \omega, \mu) \\ B_r &= \gamma_1(\omega, \mu)A - (n-1)\alpha B + \mathrm{i}BP_1\left(|A|^2, \mathrm{i}(A\overline{B} - \overline{A}B)^2, \alpha; \omega, \mu\right) \\ &\quad + AP_2\left(|A|^2, \mathrm{i}(A\overline{B} - \overline{A}B), \alpha; \omega, \mu\right) + \mathcal{R}_2(A, B, \alpha; \omega, \mu) \\ \alpha_r &= -\alpha^2. \end{aligned}$$

Here, P_1 and P_2 are complex polynomials in their arguments, and the coefficients depend smoothly on ω and μ. The full equation is invariant under reversibility acting through $r \mapsto -r$, $\alpha \mapsto -\alpha$, $B \mapsto -B$, and under time shift acting through $(A, B) \mapsto (\mathrm{e}^{\mathrm{i}\theta}A, \mathrm{e}^{\mathrm{i}\theta}B)$. The remainder terms $\mathcal{R}_j$ satisfy

$$\mathcal{R}_j(A, B, \alpha; \omega, \mu) = \mathrm{O}\left((|A| + |B|)^m + |\alpha|^m(|A| + |B|)^3\right).$$

Expanding and collecting the leading order terms gives

$$(98)\qquad \begin{aligned} A_r &= B + \mathrm{O}\left((|A| + |B|)^3\right) \\ B_r &= (-\mu_1 + \mathrm{i}\hat{\omega})\,A - (n-1)\alpha B + \gamma_2 A|A|^2 \\ &\quad + \mathrm{O}\left((|\alpha| + |\mu_1| + |\omega_1|)|A|^3 + \sum_{j=0}^{2} |A^j B^{3-j}| + (|A| + |B|)^5\right). \end{aligned}$$

Here, we have set $\gamma_1 = -\mu_1 + \mathrm{i}\hat{\omega}$ with $\mu = (\mu_1, \dots, \mu_p)^T$ and $\gamma_2 = \partial_1 P_2(0, 0, 0; 0, \omega_*)$. As in Chapter 2, Section 4, we have $\frac{\mathrm{d}\hat{\omega}}{\mathrm{d}\omega} \neq 0$ in $\omega = \omega_*$, $\mu = 0$. Scaling $A = |\mu_1|^{1/2}\tilde{A}$, $B = |\mu_1|\tilde{B}$, $r = |\mu_1|^{-1/2}\tilde{r}$, $\alpha = |\mu_1|^{1/2}\tilde{\alpha}$, and $\hat{\omega} = |\mu_1|\tilde{\omega}$ gives

$$(99)\qquad \begin{aligned} \tilde{A}_{\tilde{r}} &= \tilde{B} + \mathrm{O}\left(|\mu_1|^{1/2}\right) \\ \tilde{B}_{\tilde{r}} &= (\pm 1 + \mathrm{i}\tilde{\omega})\tilde{A} - (n-1)\tilde{\alpha}\tilde{B} + \gamma_2\tilde{A}|\tilde{A}|^2 + \mathrm{O}\left(|\mu_1|^{1/2}\right). \end{aligned}$$

If $\gamma_2 = 0$, we assume $\partial_1 P_2(0, 0; \mu, \omega_*) = \mu_2 + \mathrm{i}\mu_3$, a generic assumption up to diffeomorphic transformations in parameter space μ, ω. We then scale $\mu_2 = \tilde{\mu}_2|\mu_1|^{1/2}$, $\mu_3 = \tilde{\mu}_3|\mu_1|^{1/2}$, $A = |\mu_1|^{1/4}\tilde{A}$, $B = |\mu_1|^{3/4}\tilde{B}$, $x = |\mu_1|^{-1/2}\tilde{x}$, and $|\mu_1|\tilde{\omega} = \hat{\omega}$ to arrive at

$$(100)\qquad \begin{aligned} \tilde{A}_{\tilde{r}} &= \tilde{B} + \mathrm{O}\left(|\mu_1|^{1/2}\right) \\ \tilde{B}_{\tilde{r}} &= (\pm 1 + \mathrm{i}\tilde{\omega})\tilde{A} - (n-1)\tilde{\alpha}\tilde{B} + (\tilde{\mu}_2 + i\tilde{\mu}_3)\tilde{A}|\tilde{A}|^2 + \gamma_3\tilde{A}|\tilde{A}|^4 \\ &\quad + \mathrm{O}\left(|\mu_1|^{1/2}\right) \end{aligned}$$

with $2\gamma_3 = \partial_1^2 P_2(0,0,0;0,0)$.

4. Heteroclinics in the reduced equation

We are interested in solutions which converge for $r \to \infty$ to plane-wave solutions $U(t,r) \sim U_0(\omega t - kr)$, with U_0 being 2π-periodic. In our reduced systems (99) and (100), at $\alpha = 0$, we find these waves in the form $\tilde{A}(\omega;\mu)\mathrm{e}^{\mathrm{i}\tilde{k}\tilde{r}}$, where $\omega = \omega(\tilde{k};\mu)$. For our purposes, the most important quantity associated to these plane waves is the group velocity $c_g(k) = \frac{\mathrm{d}\omega(\tilde{k})}{\mathrm{d}\tilde{k}}$. Heteroclinics converging to these plane-wave solutions represent target patterns in the full reaction-diffusion system, where far from the center, we see concentric circles, travelling with constant speed away from the center or towards the center. Locally, far away from the center, the waves resemble plane waves since the curvature of the circle is small. If the group velocity, associated to these plane waves is positive, then perturbations are transported away from the center and we think of the center as a source for the outgoing waves. If the group velocity is negative, then information is transported towards the center and we think of the target patterns as sinks, where incoming concentric waves collide and annihilate.

In this section, we address the question of existence of target patterns in the far-field equation, on $\tilde{\mathcal{W}}^{\mathrm{c}}_{+}$, only. Persistence and matching with the center is postponed to Section 5.

The simplest example of a supercritical nondegenerate Hopf bifurcation (H) does not give rise to target patterns, where the center acts as a source for the *outgoing* waves — at least not close to the variational case of the real Ginzburg-Landau equation [**KH81b**]; see however Chapter 5, Section 3. On the other hand, shock-type defects, where concentric waves collide in the center were found in [**Gre78**]. Both studies are perturbation analyses around the trivial constant solution $\tilde{A}(\tilde{r}) \equiv 1/\sqrt{\gamma_2}$ in the limit $\operatorname{Im}\gamma_2 = 0$, $\operatorname{Re}\gamma_2 > 0$, and $\tilde{\omega} = 0$, and $\mu_1 > 0$ in (99),

$$\tilde{A}_{\tilde{r}\tilde{r}} = -\frac{n-1}{\tilde{r}}\tilde{A}_{\tilde{r}} - \tilde{A} + \operatorname{Re}\gamma_2\tilde{A}|\tilde{A}|^2.$$

We restate the existence of target sinks.

PROPOSITION 4.9. [Target sinks] [**Gre78**] *For* $\operatorname{Im}\gamma_2$ *and* $\tilde{\omega}$ *sufficiently small, there exists a solution* $\tilde{A}(\tilde{r})$ *of equation (99) with* $\mu_1 = 0$*, such that* $A(r) \sim \tilde{A}_*(\tilde{k})\mathrm{e}^{\mathrm{i}\tilde{k}\tilde{r}} = \tilde{A}(\tilde{r};\tilde{k})$*,* $\tilde{A}_* = \sqrt{(1-k^2)/\gamma_2}$*, as* $\tilde{r} \to \infty$ *is asymptotic to a plane wave* $A(r;\tilde{k})$ *with negative group velocity* $\mathrm{d}\tilde{\omega}/\mathrm{d}\tilde{k} < 0$*. The solution is a transverse heteroclinic orbit of the non-autonomous radial dynamics.*

The group velocity $\mathrm{d}\tilde{\omega}/\mathrm{d}\tilde{k}$ here is meant as the derivative of the nonlinear dispersion relation $\tilde{\omega}(\tilde{k})$. It could also be computed from the linear dispersion relation in the linearization about the plane wave [**SS00c**]. Note that there is a whole family of sinks, parameterized by the asymptotic wave number $\tilde{k}$.

The second example exhibits a mechanism for the creation of target patterns, where the asymptotic wave trains have positive group velocity, that is, where they transport perturbation away from the center. The idea is to start with a one-dimensional, $x \in \mathbb{R}$, *coexistence pattern*, see Chapter 2, Sections 3.5 and 4.3. Although the central argument is global in nature, we illustrate it in the case of the degenerate Hopf bifurcation, equation (100).

We assume that, for a fixed value of $\tilde{\mu}_2 = \tilde{\mu}_2^*$, there exists a coexistence pattern $\tilde{A}(\tilde{r};\omega)$ in the one-dimensional (asymptotic, $r \to \infty$) problem, that is a solution to (100) with $\tilde{\alpha} \equiv 0$ and $\tilde{A}(\tilde{r};\omega) \to 0$ for $r \to -\infty$ and $\tilde{A}(\tilde{r};\omega) \to \tilde{A}_* e^{i\tilde{k}_*\tilde{r}}$ for $\tilde{r} \to \infty$. In Chapter 2, Section 4, we already observed that coexistence patterns are codimension-one phenomena and proved existence in a particular parameter regime.

Furthermore, assume that the heteroclinic orbit is transverse in c and in $\tilde{\mu}_2$ as parameters. In other words, the speed c crosses zero with non vanishing speed when $\tilde{\mu}_2$ is varied about $\tilde{\mu}_2^*$; see Remark 2.34.

We call a heteroclinic orbit satisfying these condition a *transverse* coexistence pattern.

PROPOSITION 4.10. [Target sources in degenerate Hopf] *Assume that there exists a transverse coexistence pattern for the asymptotic equation, $\tilde{r} = \infty$, of (100) at $\tilde{\mu}_2 = \tilde{\mu}_2^*$. Then for all $\tilde{\mu}_2$ close to $\tilde{\mu}_2^*$, there exists a heteroclinic in the reduced, truncated system on the center manifold if, and only if, $M(\mu_2 - \mu_2^*) > 0$. The coefficient M is given as the product of the two Melnikov integrals with respect to c and $\tilde{\mu}_2$. The heteroclinics constructed are transverse in radial dynamics.*

REMARK 4.11. *The sign of the product of the two Melnikov functions is precisely the sign of the derivative $\mathrm{d}c/\mathrm{d}\tilde{\mu}_2$. In particular, the condition for existence gives coexistence patterns in the region of parameter space where c would be positive, that is, where the homogeneous state which occupies the center of the target pattern would spread into the outgoing waves in one spatial dimension. The physical interpretation can be formulated as a positive interfacial energy, preventing the homogeneous state from spreading into the outgoing waves.*

Proof. *[of Proposition 4.10]* The proof is similar to the proof of Proposition 2.24; see also Chapter 2, Figure 4. We denote by $\tilde{W}_+^{\mathrm{u}}$ the center-unstable manifold of zero in the extended phase space $(\tilde{A}, \tilde{B}, \tilde{\alpha})$ and by $\tilde{W}_+^{\mathrm{s}}$ the center-stable manifold of the plane wave $\tilde{A}_* e^{i\tilde{k}_*\tilde{r}}$. The condition on transversality in c is equivalent to transversality with respect to $\tilde{\alpha}$ since $\tilde{\alpha}' = -\tilde{\alpha}^2$ is quadratic and therefore α appears as a constant parameter in the linearization along the heteroclinic, at the same position as the wave speed c. Therefore, $\tilde{W}_+^{\mathrm{u}}$ and $\tilde{W}_+^{\mathrm{s}}$ intersect transversely in the extended phase space. The sign condition in the Proposition ensures that this intersection persists at some parameter value $\tilde{\alpha}_*(\mu_2) > 0$. By the λ-Lemma, the shooting manifold W_-^{cu} is exponentially close in $1/\tilde{\alpha}_*$ to $\tilde{W}_+^{\mathrm{u}}$ and therefore intersects $\tilde{W}_+^{\mathrm{s}}$ transversely, at a position exponentially close to $\alpha_*(\mu_2)$, in the flow-invariant subspace $\alpha > 0$. ∎

5. Persistence

We argue that the solutions in Propositions 4.9 and 4.10 persist for the full reaction-diffusion system.

THEOREM 4.12. *Consider a linearly generically unfolded Hopf instability (H) with a single parameter μ and with a cubic coefficient $\gamma_2 = 1 + \mathrm{i}\gamma_2^{\mathrm{i}}$ from (99) with positive real part, that is, periodic plane waves exist in the region where the trivial state is unstable. Then the target sinks found in the universal equation (99), Proposition 4.9, give rise to a family of target sinks for the full reaction-diffusion system in the region $\mu = \mu_1 > 0$.*

The solutions in the theorem correspond to radially symmetric, temporally periodic target sinks in the full reaction-diffusion system. They bifurcate from a homogeneous equilibrium state when it undergoes a Hopf instability (H). Existence is guaranteed for small values γ_2^{i}, which would correspond to almost equal amounts of linear and nonlinear dispersion in a complex Ginzburg-Landau equation. Phenomenologically, concentric circles of waves with large wavelength shrink towards the origin, where they collapse.

Proof. Similarly to the proof of Theorem 3.18, we have an expansion of the matching subspace W_-^{cu} for each nonzero value of α

$$W_-^{\mathrm{cu}} = \left\{(A,B) = \left(a, \psi_-^{\mathrm{u}}(a)\right); \, a \in \mathbb{C}, \, |a| \leq \delta_-\right\},$$

with $\psi_-^{\mathrm{u}}(a) = \mathrm{O}(|a|^2 + |\mu| + |\omega|)$.

By the transversality of the heteroclinic in the universal equation (99), with $\mu_1 = 0$, and by the λ-Lemma, we have, for any $\epsilon > 0$, the expansion

$$W_+^{\mathrm{s}} = \left\{(\tilde{A}, \tilde{B}) = (\mathrm{e}^{\mathrm{i}\varphi} a_*, b) + \mathrm{O}(\tilde{r}^{1-\epsilon}); \, b \in \mathbb{C}, \, |b| \leq \delta_+; \, 0 \leq \varphi < 2\pi\right\},$$

in the scaled coordinates. The constant a_* is given as the value of the target sink from Proposition 4.9 in $r = 0$. The constant δ_+ can be chosen arbitrarily large by decreasing $\tilde{r}$. In the original coordinates this gives

$$W_+^{\mathrm{s}} = \left\{(A,B) = \left(\mathrm{e}^{\mathrm{i}\varphi} a_* |\mu|^{1/2} + \mathrm{O}(|\mu|^{3/2-\epsilon}), |\mu| b\right); \, b \in \mathbb{C}, \, |b| \leq \delta_+; \, 0 \leq \varphi < 2\pi\right\}.$$

The first matching equation gives

$$a = |\mu|^{1/2} \mathrm{e}^{\mathrm{i}\varphi} a_* + \mathrm{O}(|\mu|^{3/2-\epsilon}).$$

Plugging the result into the equation for B gives

$$|\mu| b = \psi_-^{\mathrm{u}}(|\mu|^{1/2} \mathrm{e}^{\mathrm{i}\varphi} a_* + \mathrm{O}(|\mu|^{3/2-\epsilon})) = \mathrm{O}(\mu).$$

With the implicit function theorem, we obtain a unique solution $b(\varphi; \mu)$ for $\mu > 0$ small. ∎

The persistence of coexistence patterns, Proposition 4.10, is conceptually simpler. By hyperbolicity of the zero-solution, the matching subspace W_-^{cu} is transverse to the center-stable manifold W_+^{s} of the origin and we can repeat the arguments from the proof of Proposition 4.10 to find

THEOREM 4.13. *Consider a linearly generically unfolded Hopf instability (H) with degenerate cubic coefficient $\gamma_2 = 0$. The target sources found in the universal equation, Proposition 4.10, give rise to a family of target sources for the full reaction-diffusion system. In particular, target sources exist in a region in parameter space with nonempty interior.*

We did not make the conditions which guarantee the existence of target sources for the full reaction-diffusion system explicit. The situation we unfolded was codimension 4, with vanishing μ at the instability threshold, vanishing cubic terms

$\gamma_2 \in \mathbb{C}$ and zero imaginary part of the quintic coefficient. These codimension-4 points in the space of reaction-diffusion systems lies in the closure of reaction-diffusion systems which possess target sources, that is, solutions which in the far-field consist of equidistantly spaced concentric circles, propagating away from the center with constant, positive phase and group velocity.

CHAPTER 5

Discussion

The mere existence of radially symmetric solutions, the major topic of this work, has provided us with some of insight into the dynamics of reaction-diffusion systems on $\mathbb{R}^n$. Nevertheless, the general method presented here, might well serve as a tool for more delicate questions in this context. We outline several extensions and point out limitations of our approach. As a first topic, we discuss stability, in particular spectral bifurcation problems, Section 1. We then point out how to find non radially symmetric patterns, Section 2, exhibiting in particular limitations of the approach via radial dynamics. In the last part, Section 3, we discuss how a small hole might affect the bifurcation analysis. Emphasis is laid on the existence of target sources for certain Robin boundary conditions. We conclude with a short summary, directing us towards a number of open questions.

1. Stability

We illustrate a typical stability analysis in showing instability of patterns in two examples. First, we characterize the critical spectrum of localized humps in an instability of type (O), found in Theorem 3.18; see Section 1.1. We then argue that coexistence patterns, like the ones found in Theorems 3.19 and 4.13, typically possess one unstable eigenvalue, Section 1.2. The associated eigenfunction mimics radial motion of the interface between the two stable states.

For the stability analysis of some of the one-dimensional patterns, we refer to [**GL**]. In case (O), the spectra of bifurcating, stationary solutions can actually be computed from a scaled, reduced problem. We find the stability properties known from scalar reaction-diffusion systems. In particular, only monotone solutions, namely the coexistence patterns and the travelling waves in case of a cusp, Proposition 2.24 and Remark 2.25, are stable. Spatially periodic solutions, even those, which are close to the coexistence patterns in the phase space of the reduced system, Chapter 2, Figure 4, are unstable due to continuous spectrum in the right half plane.

We conjecture that there are stable *and* unstable coexistence patterns in case of a weakly subcritical Turing instability, Chapter 2, Section 3.5. Neglecting nonadiabatic effects, we have a circle of heteroclinics, parameterized by complex phase. Within the approximation of the cubic-quintic Ginzburg-Landau equation, this circle is stable, since it is actually a local minimizer of the Ginzburg-Landau energy

$$J[A] = \int_x \left(\frac{1}{2}|A_x|^2 + \frac{1}{2}|A|^2 + \frac{1}{4}\nu_{\mathrm{Maxw}}|A|^4 + \frac{1}{6}|A|^6 \right) \mathrm{d}x.$$

Adding nonadiabatic effects, we expect that the circle breaks up, with two remaining heteroclinics, one of which we expect to be stable, the other unstable.

Stability questions for the complex cubic-quintic Ginzburg-Landau equation are largely open. We mention [**KS98**], [**vSH92**], [**KR00**] and the references therein, for some results in this direction. Nevertheless, we expect the coexistence patterns, found in Chapter 2, Section 4.3, and the travelling waves described in Remark 2.34, to be stable, as perturbations from stable coexistence patterns in the variational case of real cubic and quintic coefficients. The general procedure for a rigorous stability analysis of time-periodic patterns has been outlined in [**SS00c, SS00d**]. In particular, group velocities of the travelling waves play a crucial role for the location of the spectrum.

We return to radially symmetric patterns and explain in more detail the main ideas. The key to a stability analysis is a description of the spectrum of the linearization. Given the bifurcating, radially symmetric solution $U_*(r;\mu)$, we investigate the linearized operator

$$\mathcal{L}_\mu^* U = D\triangle U + \partial_U F(U_*(r;\mu);\mu)U.$$

We consider $\mathcal{L}_\mu^*$ on the space of square-integrable functions $L^2(\mathbb{R}^n,\mathbb{R}^N)$, with domain of definition $H^2(\mathbb{R}^n,\mathbb{R}^N)$. We say, a complex number λ belongs to the *spectrum* $\mathrm{spec}\,(\mathcal{L}_\mu^*)$, if $\lambda-\mathcal{L}_\mu^*$ does not possess a bounded inverse. We define the *point spectrum* $\mathrm{spec}_{\mathrm{point}}\,(\mathcal{L}_\mu^*)\subset\mathrm{spec}\,(\mathcal{L}_\mu^*)$ as the subset of $\lambda\in\mathbb{C}$, where $\lambda-\mathcal{L}_\mu^*$ is Fredholm with index zero, that is, the range is closed and the dimensions of kernel and cokernel coincide. The complement $\mathrm{spec}_{\mathrm{ess}}\,(\mathcal{L}_\mu^*):=\mathrm{spec}\,(\mathcal{L}_\mu^*)\setminus\mathrm{spec}_{\mathrm{point}}\,(\mathcal{L}_\mu^*)$ is called the *essential spectrum*. By robustness of Fredholm properties and indices under compact perturbations, we have the following lemma.

LEMMA 5.1. *Assume that* $U_*(r;\mu)\to U_\infty(\mu)$ *as* $r\to\infty$. *Then*

$$\mathrm{spec}_{\mathrm{ess}}\,(\mathcal{L}_\mu^*)=\mathrm{spec}_{\mathrm{ess}}\,(\mathcal{L}_\mu^\infty),$$

with

$$\mathcal{L}_\mu^\infty U = D\triangle U + \partial_U F(U_\infty(\mu);\mu)U.$$

Furthermore,

$$\mathrm{spec}_{\mathrm{ess}}\,(\mathcal{L}_\mu^\infty)=\left\{\lambda\,|\,\det\left(-Dk^2+\partial_U F(U_\infty(\mu);\mu)-\lambda\right)\right)=0 \text{ for some } k\in\mathbb{R}\right\}.$$

Although, in particular situations, it might turn out to be difficult to actually compute the essential spectrum, the analytically harder problem is to locate and track the point spectrum.

LEMMA 5.2. [**SS00d**, Lem. 5.7] *Assume that* $\lambda\in\mathrm{spec}_{\mathrm{point}}\,(\mathcal{L}_\mu^*)$. *Then there exists a smooth eigenfunction* $U_{\mathrm{e}}(x)$, *which is exponentially localized* $|U(r)|\le \mathrm{e}^{-\eta|x|}$ *for some* $\eta>0$.

In particular, λ belongs to the point spectrum if, and only if, there exists an eigenfunction in L^2. Restricting to $x\in\mathbb{R}^2$ and introducing polar coordinates (r,φ) and angular Fourier decomposition, we find that the point spectrum consists precisely of those values of λ, for which there exists a $k\in\mathbb{R}$ and a bounded solution $U_{\mathrm{e}}(r)$ to

$$D(U''+\frac{1}{r}U'-\frac{k^2}{r^2}U)+\partial_U F(U_*(r;\mu);\mu)U=\lambda U.$$

For $\mu=0$, $U_*(r;\mu)\equiv 0$ and the spectrum consists entirely of essential spectrum. The heart of the spectral analysis now consists in the location of eigenvalues popping out of the essential spectrum for $\mu>0$ small.

1.1. Instability in the fold. We illustrate the bifurcation problem for the eigenvalues in the case of a fold in two spatial dimensions, $n = 2$. Recall from Theorem 3.18, that for $\mu > 0$, say, a stationary, radially symmetric, localized solution $U_*(r;\mu) = \mathrm{O}(\mu^{1/2})$ of the reaction-diffusion system (2) bifurcates from the trivial solution $U \equiv 0$. Using perturbation arguments for the linearization about the trivial state, we can conclude that $\mathrm{Re}\,\lambda \leq C\mu^{1/2}$ for some constant C and all $\lambda \in \mathrm{spec}\,(\mathcal{L}_\mu^*)$. From Lemma 5.1, it is not hard to conclude that the essential spectrum is strictly contained in the open left half plane. We still have to locate eigenfunctions. We look for eigenfunctions as bounded solutions to

$$
\begin{aligned}
U' &= V \\
V' &= -\frac{1}{r}V + \frac{k^2}{r^2}U + D^{-1}\left(-\partial_U F(U_*(r;\mu);\mu)U + \lambda U\right).
\end{aligned}
\tag{101}
$$

This linear differential equation can be reduced, to a center manifold $\tilde{\mathcal{W}}_+^{\mathrm{c}}$ in the same fashion as the nonlinear problem in Theorem 3.7. Since the equation is linear, we do not need cut-off functions. We include the eigenvalue parameter λ as an additional parameter. Since the equation is analytic in λ, the asymptotic center manifold $\tilde{\mathcal{W}}_+^{\mathrm{c}}$ is also analytic in λ. Substituting $\mu = \epsilon^2$, we find that the equation and the invariant manifolds are smooth in ϵ. In the construction of the matching manifolds $\tilde{\mathcal{W}}_+^{\mathrm{c}}$, we exploit the fact, that in $\epsilon = 0$, $\lambda = 0$, and $k = 0$, equation (101) coincides with the linearization of the nonlinear problem, equation (46). For $k \neq 0$, the construction of $\tilde{\mathcal{W}}_-^{\mathrm{cu}}$ requires a slight modification. In logarithmic spatial time $\tau = \log r$, we find

$$
\begin{aligned}
U_\tau &= W \\
W_\tau &= k^2 U + r^2 D^{-1}\left(-\partial_U F(U_*(r;\mu);\mu)U + \lambda U\right) \\
r_\tau &= r.
\end{aligned}
$$

Here, the equilibrium $U = W = 0$, $r = 0$, is hyperbolic with $N+1$-dimensional unstable manifold $\tilde{\mathcal{W}}_-^{\mathrm{u}}$, which contains precisely the solutions which are bounded, actually $\mathrm{O}(r^k)$, as $r \to 0$. We use this unstable manifold instead of $\tilde{\mathcal{W}}_-^{\mathrm{cu}}$ in Proposition 3.5. Next, the center-stable E_+^{cs} and center eigenspaces E_+^{c} at $r = +\infty$ do not depend on k, and the construction of $\tilde{\mathcal{W}}_+^{\mathrm{cs}}$, Proposition 3.4 and $\tilde{\mathcal{W}}_+^{\mathrm{c}}$ is the same for all k. Perturbation arguments as in the proof of Theorems 3.3 and 3.7 then finish the construction of the asymptotic center manifold $\tilde{\mathcal{W}}_+^{\mathrm{c}}$.

On the center manifold, we introduce coordinates (A, B) as in Lemma 3.9. Rescaling the reduced equation on $\tilde{\mathcal{W}}_+^{\mathrm{c}}$ according to $r = \mu^{-1/4}\tilde{r}$, $\lambda = \mu^{1/2}\tilde{\lambda}$, $U_* = \mu^{1/2}\tilde{U}_*$, $A = \mu^{1/2}\tilde{A}$, and $B = \mu^{3/4}\tilde{B}$, we find to leading order the linearization of the universal equation

$$
\begin{aligned}
\tilde{A}_{\tilde{r}} &= \tilde{B} \\
\tilde{B}_{\tilde{r}} &= -\frac{1}{\tilde{r}}\tilde{B} + \frac{k^2}{\tilde{r}^2}\tilde{A} - 2\tilde{U}_*\tilde{A} + \gamma_L\tilde{\lambda}\tilde{A} + \mathrm{O}\left(\mu^{1/2}(|\tilde{A}| + |\tilde{B}|)\right),
\end{aligned}
\tag{102}
$$

with the additional terms $\gamma_L\tilde{\lambda}\tilde{A}$ accounting for the temporal eigenvalue λ, and $\frac{k^2}{\tilde{r}^2}\tilde{A}$ for the possible angular dependence of eigenfunctions. The coefficient γ_L is obtained by projecting $(0, D^{-1}U_0)^T$ on the center eigenspace, spanned by $\{(U_0, 0)^T, (0, U_0)^T\}$ along the spectral complement. Here, U_0 is a vector in the kernel of $\partial_U F(0;0)$; see also Lemma 2.11. With U_0^* as the vector in the kernel of the adjoint $\partial_U F^*(0;0)$

and normalization $(U_0, U_0^*) = 1$, we obtain

$$\gamma_L = (U_0^*, D^{-1}U_0) > 0,$$

since $D > 0$ is positive.

If we set $\mu = 0$ in (102), we find a universal equation, which is simply the linearization about the single hump in the equation $u_t = \Delta u - 1 + u^2$. In particular, the only bounded solutions for values of $\tilde{\lambda}$ in the right complex half plane occur for

- $k = 0$, $\tilde{\lambda} = \lambda_0 > 0$ with positive eigenfunction;
- $k = \pm 1$, $\tilde{\lambda} = 0$ from translation of the pattern, a double eigenvalue.

More generally, (102) defines for each fixed k a self-adjoint Sturm-Liouville eigenvalue problem on the unbounded half line $\tilde{r} > 0$. Eigenfunctions above the essential spectrum are simple and ordered by the number of zeroes.

The persistence problem for $k = 0$ is similar to Theorem 3.18. The matching problem gives a linear equation with coefficients, which are analytic in λ and smooth in $\sqrt{\mu}$. For λ not in the spectrum of the universal, scaled equation, we find the unique, trivial zero-solution. Close to $\lambda = \sqrt{\mu}\lambda_0$, we find nontrivial solutions to leading order as zeroes of the determinant of the linear operator in the matching problem. Actually, simple eigenvalues correspond to simple zeroes and persist by the implicit function theorem.

For $k \neq 0$, the leading eigenfunctions are generated by translation and the eigenvalue $\lambda = 0$ is an eigenvalue for the full system. More generally, the persistence problem can be treated as in the case $k = 0$, only using slightly different coordinates $r = \mathrm{e}^\tau$, $A_\tau = B$, near $r = 0$:

$$A_\tau = B, \quad B_\tau = k^2 A + \mathrm{O}(r^2), \quad r_\tau = r.$$

The origin $A = B = 0$, $r = 0$ is hyperbolic and we find expansions for the matching subspace W_-^{cu} of the form

$$W_-^{\mathrm{cu}}(r) = W_-^{\mathrm{u}}(r) = \left\{(A, B) = (a, ka) + \mathrm{O}(r^2);\ a \in \mathbb{R}\right\}.$$

The matching procedure with the so-defined matching subspace W_-^{cu} is the same as in the case of $k = 0$. The simple eigenfunctions persist as nontrivial, transverse intersections of the matching subspaces in the parameter λ.

1.2. Instability of coexistence patterns. Beyond local bifurcations, we have seen that coexistence in one space dimension for a specific parameter value typically (two Melnikov integrals are supposed to be nonzero) gives coexistence patterns in higher space dimension for an open set of parameter values. The effect was illustrated in case of a cusp, where stable, homogeneous equilibria may coexist, and in case of a degenerate Hopf bifurcation, where stable homogeneous equilibria may coexist with wave-trains, travelling away from the interface. We argued that interfacial energy balanced the energetic difference between the two different states, which is responsible for the nonzero speed of propagation for a detuned parameter value in one space dimension. The argument suggests that coexistence patterns constructed in this fashion are unstable. We outline here, how to characterize this instability mechanism as an isolated, unstable eigenvalue in a bifurcation analysis.

For convenience, consider the case of a stationary, homogeneous instability (O), with a cusp in the kinetics, Theorem 3.19. In the extended phase space, multi-dimensional coexistence patterns arise in a heteroclinic bifurcation; see Chapter 3, Figure 3. Most of the arguments are similar in case of a degenerate Hopf bifurcation.

The proof actually does not exploit smallness of the patterns and applies more generally to coexistence patterns in reaction-diffusion systems.

Emphasizing the general nature of this bifurcation, we introduce a short notation. Denote by $z \in [0,1]$ the compactified radial "time variable", with $z = 1 - 1/r$ for $r \to \infty$ and $z = r$ for $r \to 0$. Also, let Y denote the state variables (A, B) near $r = \infty$ and (rA, rB) near $r = 0$ and let $\tilde{Y} = (Y, z)$. Radial time is scaled like $r = \mathrm{e}^{\tau}$ near $r = 0$ and $r = \tau$ near $r = \infty$. The dynamics can then be written in the compact for $\tilde{Y}' = G(\tilde{Y})$.

At bifurcation we have two heteroclinic orbits in the extended phase space $q_0(\tau)$ and $q_1(\tau)$, with

$$q_0(\tau) \to p_0, \text{ for } \tau \to -\infty, \quad q_0(\tau) \to p_1, \text{ for } \tau \to +\infty$$

and

$$q_1(\tau) \to p_1, \text{ for } \tau \to -\infty, \quad q_1(\tau) \to p_2, \text{ for } \tau \to +\infty$$

with $p_0 = 0$, $p_1 = (Y = 0, z = 1)$, and $p_2 = (p_y, 1)$. Here, p_0 denotes the center of the multi-dimensional pattern, p_1 denotes the inner state in the coexistence pattern, and p_y denotes the outer stable state. The first heteroclinic is actually the trivial connection $Y = 0$ from $z = 0$ to $z = 1$. The second heteroclinic q_1 denotes the one-dimensional coexistence pattern, as found in Proposition 2.24, for example. For nearby parameter values, we find heteroclinics $q(\tau)$ connecting p_0 to p_2, see Proposition 3.16 and Theorem 3.19.

We address stability of the heteroclinics $q(\tau)$, next. For the linearized stability of q_0 and q_1, we linearize the equation about these heteroclinics, including the spectral parameter λ. We find two linear equations for $\tilde{Y}$

$$\tilde{Y}' = G'(q_0(\tau); \lambda)Y, \quad \text{and} \quad \tilde{Y}' = G'(q_1(\tau); \lambda)Y. \tag{103}$$

The first equation in (103) does not possess nontrivial bounded solutions for λ close to zero since the first state was assumed to be stable. The second equation possesses a bounded solution for $\lambda = 0$ which is induced by spatial translation of the coexistence profile. We are interested in nontrivial bounded solutions for the linearization about $q(\tau)$

$$\tilde{Y}' = G'(q(\tau); \lambda)Y. \tag{104}$$

In particular, we would like to recover a radially symmetric eigenfunction to an eigenvalue λ close to zero, which resembles the translational eigenfunction, $\lambda = 0$ of q_1.

A similar problem was studied in [**SS00a**]. Given the algebraic multiplicities ℓ_1 and ℓ_2 of a possible eigenvalue λ for the individual heteroclinics q_0 and q_1, it is shown there, that the sum of the algebraic multiplicities to all eigenvalues λ' for the linearization (104) about the coexistence profile $q(\tau)$ in a sufficiently small neighborhood of λ is the sum $\ell_1 + \ell_2$.

For the problem under consideration here, this implies that the zero eigenvalue corresponding to translation of the interface continues to a unique simple eigenvalue with radially symmetric eigenfunction in a neighborhood of zero for the coexistence pattern. The motion in the direction of this slow eigenspace corresponds to growing or shrinking of the region occupied by p_0, when the initial interface is perturbed.

The interesting question now is, whether this unique critical eigenvalue is actually stable or unstable. A more detailed perturbation analysis shows that, in accordance with the physical interpretation, the eigenvalue is always unstable, here; see also Remark 4.11.

The analysis can be extended to oscillatory coexistence patterns. the translation eigenvalue of the asymptotic coexistence pattern is embedded in the essential spectrum, but not contained in the *absolute spectrum* due to the nonvanishing group velocity of the outgoing periodic waves; see [**SS00b, SS00c, SS00a**] for the definition of the absolute spectrum, spectra of time-periodic patterns, and the relation to group velocities.

2. Beyond radial symmetry

Radially symmetric patterns are the only localized solutions for a large class of *scalar* elliptic equations [**GNN97**]. Proofs of this fact strongly exploit comparison principles. The bifurcation results for elliptic *systems* presented here suggest that in the case of a one-dimensional kernel, case (O), the scalar structure is recovered. However, the uniqueness of the radially symmetric solutions from Theorem 3.18 in the class of localized, not necessarily radially symmetric, functions remains an open problem.

Dropping the assumption of spatial localization, many other patterns are found, starting with solutions depending only on one spatial coordinate x_1; see Chapter 2. More interestingly, periodic arrangements of single hump patterns along lattices in $\mathbb{R}^n$ can be found, following the arguments in [**PSS97, LPSS00**]. We outline the strategy for two spatial dimensions $\underline{x} = (x, y)$. Suppose we are given a stationary, localized, radially symmetric solution $U(r)$ to our reaction-diffusion system. We may then write the elliptic system as an ill-posed dynamical system in the x-coordinate

$$U_x = V, \quad V_x = -\Delta_y U + D^{-1}F(U)$$

on the phase space $(U, V)(\cdot) \in H^1(\mathbb{R}, \mathbb{R}^N) \times L^2(\mathbb{R}, \mathbb{R}^N)$. The radially symmetric solution $U(r)$ becomes a homoclinic orbit in x-dynamics. By rotational invariance, it is actually a reversible homoclinic, that is, it is set-wise invariant under the involution $(U, V) \mapsto (U, -V)$. Restricting to even functions of y factors out the translational symmetry in the y-direction. If the generalized kernel of the linearization about the single-hump consists precisely of x- and y-derivatives (as is the case for the localized solutions from Theorems 3.18 and 3.19), the homoclinic is nondegenerate in the sense of [**LPSS00**, Hypothesis H4,(ii)]. A proof of this fact, exhibiting the general relation between transversality (here, the nondegeneracy condition) and the generalized kernel, can be found in [**SS00e**, Prop. 5.2]. From [**LPSS00**, Cor. 4], we may then conclude that the homoclinic is accompanied by periodic patterns $U_{L_x}(x, y)$ with sufficiently large period $L_x \geq L_x^*$ in x [**LPSS00**, Cor. 4]. On an interval of periodicity, the periodic patterns resemble the original pulse up to a correction, which is exponentially small in the wavelength L_x.

We may now consider these x-periodic patterns as homoclinic orbits in y-dynamics

$$U_y = V, \quad V_y = -\Delta_x U + D^{-1}F(U)$$

on even, L-periodic functions in x. Again, the homoclinics are nondegenerate and accompanied by y-periodic patterns, with large period $L_y \geq L_y^*(L_x)$. A refinement of the above arguments shows that actually L_y^* can be chosen independently of L_x large enough.

Summarizing, we find spatially periodic patterns $U(x, y) = U(x + L_x, y) = U(x, L_y + y)$ for all sufficiently large periods L_x, L_y; see Figure 1.

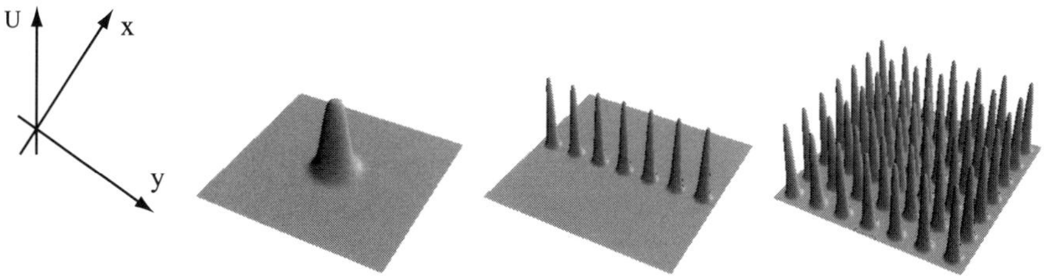

FIGURE 1. *Localized solution, x-periodic, and x-y-periodic solutions constructed in spatial dynamics.*

Since the bifurcating single hump pattern is already unstable, these patterns are also highly unstable in the full system. An analysis as in [**SS00d**] shows that there is a two-torus of spectrum $\lambda(\gamma_x, \gamma_y)$ with $\gamma_x, \gamma_y \in S^1$ exponentially close to the unstable eigenvalue λ_0^* of the single-hump pattern. However, reaction-diffusion systems may also possess localized, stable solutions; see [**Ni98, Tak97**]. It is then of high interest to determine the location of the two-torus associated to the translation eigenvalue, because it decides on stability of the periodic pattern. In both cases, the results in [**SS00d**] give formulas for the asymptotic location of this two-torus, depending on spectral and scattering-type information of the single hump pattern. We remark that the approach of considering dynamics on a spatial variable transverse to a primary homoclinic has been exploited earlier in [**HK95, HK96**] to find transverse modulation of a solution U of an elliptic equation depending only on one spatial variable.

3. Boundaries and holes

In a domain $\Omega = |x| \geq r_*$, the bifurcation theory is very similar in spirit. Instead of solving a matching problem, we now have to match the stable subspace W_+^{s} with the subspace imposed by the boundary conditions. Neumann boundary conditions imitate best the entire plane and the bifurcation results in case (O) and (H) hold in this situation. It is then interesting to see how the matching condition gets violated when we homotope the boundary conditions from Neumann to Dirichlet according to $(\cos s)U_r + (\sin s)U = 0$. For s close to 0 we have the same result as for Neumann boundary conditions. At some value of s, for Robin boundary conditions, a bifurcation occurs because transversality between the subspace imposed by the boundary conditions and W_+^{s} is violated. The bifurcation is typically a fold in case (O).

Another interesting problem is the stability of the pattern in Ω. In particular, the eigenvalues induced by the translational symmetry will typically leave the origin. Stable and unstable patterns corresponding to a pattern which is pinned to the hole, or which is repelled by the hole are possible: the saddle-node bifurcations, described above, mark the transitions between these two types of patterns.

On the other hand, boundary conditions may generate patterns which do not exist in $\mathbb{R}^n$. Target patterns, that is, time-periodic, radially symmetric patterns, which resemble concentric waves travelling away from the center, are one typical example. In [**Hag81**], a formal analysis shows that local inhomogeneities may create target patterns.

We outline how to find target sources in the complement of a disc of diameter r_*. We impose Robin boundary condition $U_r = sU$, $s > 0$, $s \ll 1$, in $r = r_* > 0$. We start with the complex Ginzburg-Landau equation, that we found in Chapter 4, Section 3 as the universal, reduced equation for the far-field shape of the patterns:

$$A' = B, \quad B' = -\frac{1}{r}B - (1 + \mathrm{i}\omega)A + (1 + \mathrm{i}\gamma)A|A|^2.$$

The constant γ was computed as the cubic normal form coefficient of the reduced vector field. Choosing the free frequency parameter $\omega = \gamma$, we find the r-independent solution $A = 1$, $B = 0$. In $r = \infty$, the solution undergoes a reversible, $\mathcal{SO}(2)$-equivariant, saddle-node bifurcation upon decreasing ω, where γ is fixed. Associated with this bifurcation, we have a three-dimensional center manifold in a neighborhood of $r = \infty$, $A = 1$, $B = 0$. Apart from the $\alpha = 1/r$-direction, we have the direction of the group orbit $(\mathrm{i}, 0)$ and the direction of the bifurcation $(0, \mathrm{i})$. In the saddle-node, two *relative* equilibria $A(r) = R\mathrm{e}^{\mathrm{i}kr}$, $R > 0$, are created with $R^2 = 1 - k^2$, $\omega = \gamma R^2$. The linearization in $r = \infty$ about the bifurcating relative equilibria possesses one stable and one unstable eigenvalue, the trivial zero eigenvalue corresponding to complex rotational equivariance, and an eigenvalue close to zero, associated with the bifurcation, whose sign is actually the sign of the group velocity. Arguing as in Proposition 3.11 and Remark 3.12, one can show that the part of the center manifold in the $\alpha = 0$-subspace is smoothly fibered in the direction of time α. The bifurcation associated with varying ω is illustrated in Figure 2.

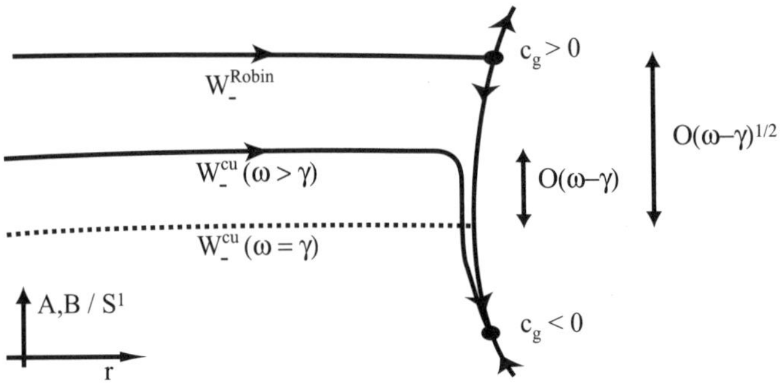

FIGURE 2. *The asymptotic center manifold W_+^{s} is plotted, with the free action of the rotations $\mathcal{SO}(2)$ factored out. The matching subspaces W_-^{cu} depending on ω and the transported boundary conditions W_-^{Robin} are shown relative to the equilibria in the $\alpha = 0$-subspace.*

An explicit computation in terms of Bessel functions shows that the one-dimensional strong stable manifold of the $k = 0$, $\omega = \gamma$-equilibrium $A = 1$, $B = 0$

on the center manifold crosses the two-dimensional matching manifold $W_-^{\rm cu}$ transversely upon varying ω. On the other hand, $W_-^{\rm cu}$ transversely intersects the center-stable manifold $W_+^{\rm cs}$ of $A = 1$, $B = 0$. For $\omega \neq 0$, we therefore obtain unique intersections of $W_-^{\rm cu}$ with strong stable fibers within $W_+^{\rm cs}$. However, the relative equilibrium with positive group velocity is unstable within the asymptotic center manifold and at distance $\mathrm{O}(\sqrt{\gamma - \omega})$, whereas the strong stable fiber, connecting to $W_-^{\rm cu}$ is based at the much smaller distance $\mathrm{O}(\gamma - \omega)$. This explains the absence of target sources in the full plane. Now, changing the boundary condition varies the intersection base point of the fiber in $W_+^{\rm cs}$ which intersects $W_-^{\rm cu}$. If this base point lies on the part of the center manifold with $c_g > 0$, then varying ω will make this base point become the relative equilibrium with positive group velocity for a certain value $\omega = \omega_*$, again, since the equilibria bifurcate with amplitude $\mathrm{O}(\sqrt{\gamma - \omega})$.

In order to determine, which sign of s gives rise to target sources, that is, positive velocity of the base point as described above, we have to investigate the linearized equation about the solution $A(r) = 1$, which reads

$$A'' + \frac{1}{r}A' - (1 + \mathrm{i}\omega)(A + \bar{A}) = 0.$$

In this linearized equation, the imaginary subspace is flow-invariant with solutions $A(r) = \mathrm{i}$ and $A(r) = \mathrm{i}/r$. In particular, a positive sign of A' at $r = r_*$ gives a positive sign for $r \to \infty$, that is, positive group velocity. With Robin boundary conditions, at $A \sim 1$ in $r = r_*$, $A' = sA \sim s$ is positive if s is positive. The group velocity is given by $2\gamma k$. Then $s\gamma > 0$ is the sign condition for Robin boundary conditions to generate target sources.

Persistence of the solutions for the full reaction-diffusion system can be shown as in Theorem 4.12.

4. Concluding remarks

We have presented a method for analyzing pattern-formation in reaction-diffusion systems in $\mathbb{R}^n$ caused by an instability of a spatially homogeneous equilibrium. The inherent complications in the analysis, caused by the essential spectrum, could be resolved by restricting to patterns, which are

- stationary or time-periodic *and*
- radially symmetric.

The strategy was to first derive an ordinary differential equation for the far-field pattern, describing long-wavelength modulations of possible patterns including weak curvature effects, represented by the nonautonomous dependence of the equation on powers of $1/r$; see Theorems 3.7, 4.8. In a second step, solutions of this reduced ordinary differential equation were matched with the inner, core region of a possible pattern.

Curvature effects do not differ between far-field expansion and the core region in case of instabilities with zero wave number (O) and (H), where the matching condition in the core region reduces to a Neumann boundary condition. However, in the case of a Turing instability, curvature plays a crucial role: in the far-field, averaging over the periodic structures qualitatively changes curvature effects and implies spatial decay of the amplitude proportional to $\sqrt{r}$ towards the center of the pattern, whereas in the core region, we find almost constant amplitude. Transversality in the construction of the defect is very weak and might prove responsible for the weak stability of these patterns.

Stability considerations remain very incomplete. Coexistence patterns as found here are unstable. One may, nevertheless ask:

Do there exist stable coexistence patterns?

As part of this question, we would like to find stable, localized solutions in a "generic", local bifurcation. It would then be of interest, how the physical reasoning of interfacial energy balancing potential difference between two different states would fail.

Also, stability proofs for temporally oscillatory patterns such as target patterns or spiral waves would help a lot in understanding this type of defects. The (formal) arguments in [**Hag82**] seem to be the only — partial — answer in this direction.

Finally, we comment on most desirable extensions of our approach of radial dynamics to non radially symmetric patterns. A positive result in this direction was presented in [**Sche98**], where the case of a Hopf instability (H) was analyzed in radial dynamics. Patterns were allowed to depend on the angular variable φ, in contrast to periodic time-dependence, as considered here. In particular, the approach allowed for a proof of existence of spiral waves, which are stationary patterns in a *corotating frame*, substituting $\psi = \varphi - \omega_{\mathrm{rot}} t$.

However, the limitations of radial dynamics become obvious upon inspection of the lattice-like configuration of single-humps, Section 2. Figure 3 shows how, in polar coordinates, the pattern $U(r, \cdot)$ fails to converge as $r \to \infty$.

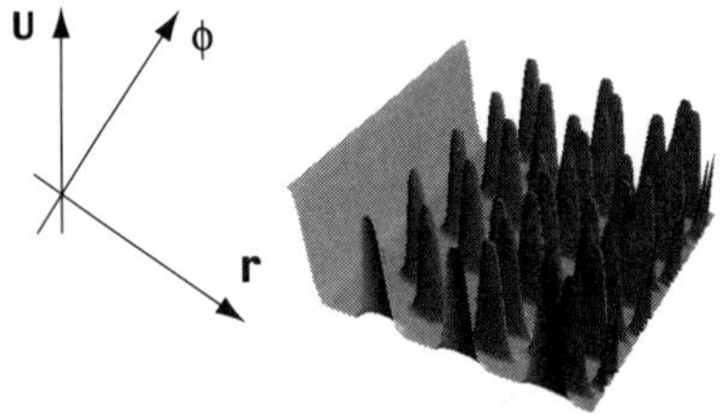

FIGURE 3. *The x-y-periodic solutions from Figure 1 plotted in polar coordinates. For $r \to \infty$, the φ-dependent patterns on $r \equiv const$ develop an increasing number of oscillations.*

Methods from dynamical systems have proven to be most efficient for qualitative descriptions of patterns in situations with one distinguished spatial direction, identified as spatial time. We have added radial dynamics as a general tool to bifurcation theory in spatially extended systems. The existence result on focus patterns in the Turing instability, as stated in Theorem 3.20, provides insight in the local structure of a particular defect in a periodically structured medium. Experiments in Rayleigh-Bénard convection [**Cro89**] exhibit many other defects and, although we may think of different "time-actions" on $\mathbb{R}^n$, we still seem to lack a systematic method to approach these patterns.

Bibliography

[AR] R. Abraham and J. Robbin. *Transversal Mappings and Flows.* Benjamin Inc., Amsterdam, 1967.

[AW30] A.A. Andronov and A. Witt. *Sur la théorie mathématique des autooscillations.* C.R. Acad. Sci. Paris **190** (1930), 256–258.

[AGRR90] F.T. Arecchi, G. Giacomelli, P.L. Ramazza, and S. Residori. *Experimental evidence of chaotic itinerancy and spatiotemporal chaos in optics.* Phys. Rev. Letters **65** (1990), 2531–2534.

[Arn83] V.I. Arnol'd. *Geometrical Methods in the Theory of Ordinary Differential Equation.* Springer-Verlag, New York, 1983.

[BLP81] H. Berestycki, P.-L. Lions, and L.A. Peletier. *An ODE approach to the existence of positive solutions for semilinear problems on* $\mathbb{R}^N$. Indiana Univ. Math. J. **30** (1981), 141–157.

[BPA00] E. Bodenschatz, W. Pesch, and G. Ahlers. *Recent Developments in Rayleigh-Bénard Convection.* Annu. Rev. Fluid Mech. **32** (2000) , 709–778.

[BBLPPTW91] M. Brambilla, F. Battipede, L.A. Lugiato, V. Penna, F. Prati, C. Tamm, and C. O. Weiss. *Transverse laser patterns. I. Phase singularity crystals.* Phys. Rev. A **43** (1991), 5090–5113.

[Bre99] H. Brezis. *Symmetry in nonlinear PDE's.* In Giaquinta, M. (ed.) et al., Differential equations: La Pietra 1996. Conference on differential equations marking the 70th birthdays of Peter Lax and Louis Nirenberg, Villa La Pietra, Florence, Italy, July 3–7, 1996. Proc. Symp. Pure Math. **65** (1999), 1–12.

[BK92] J. Bricmont and A. Kupiainen. *Renormalization group and Ginzburg-Landau equation.* Comm. Math. Physics **150** (1992), 193–208.

[BS78] S.N. Brown and K. Stewartson. *On finite amplitude Bénard convection in a cylindical container.* Proc. Roy. Soc. London A **360** (1978), 455–469.

[CF87] G. Caginalp and P.C. Fife. *Elliptic problems involving phase boundaries satisfying a curvature condition.* IMA J. Appl. Math. **38** (1987), 195–217.

[CI94] P. Chossat and G. Iooss. *The Couette-Taylor Problem.* Applied Mathematical Sciences **102**, Springer-Verlag, New York, 1994.

[CE92] P. Collet and J.-P. Eckmann. *Solutions without phase-slip for the Ginsburg-Landau equation.* Comm. Math. Physics **145** (1992), 345–356.

[Cop78] W.A. Coppel. *Dichotomies in stability theory.* Lecture Notes in Mathematics **629**, Springer-Verlag, Berlin, 1978.

[Cro89] V. Croquette. *Convective pattern dynamics at low Prandtl number: Part I & II.* Contemp. Phys. **30** (1989), 113–133 & 153–171.

[CH93] M.C. Cross and P.C. Hohenberg. *Pattern formation outside equilibrium.* Rev. Modern Phys. **65** (1993), 851–1112.

[CS90] R. Cushman and J.A. Sanders. *A survey of invariant theory applied to normal forms of vectorfields with nilpotent linear part.* Invariant theory and tableaux, Proc. Workshop, Minneapolis/MN (USA) 1988, IMA Vol. Math. Appl. **19** (1990), 82–106

[Doe96] A. Doelman. *Breaking the hidden symmetry in the Ginzburg-Landau equation.* Physica D **97** (1996), 398–428.

[DSSS00] A. Doelman, B. Sandstede, A. Scheel, and G. Schneider. In preparation.

[EI89] W. Eckhaus and G. Iooss. *Strong selection or rejection of spatially periodic patternd in degenerate bifurcations.* Physica D **39** (1989), 124–146.

[EG93] J.-P. Eckmann and Th. Gallay. *Front solutions for the Ginzburg-Landau equation.* Comm. Math. Physics **152** (1993), 221–248.

[ETBCI87] C. Elphick, E. Tirapegui, M.E. Brachet, P. Coullet, and G. Iooss. *A simple global characterization for normal forms of singular vector fields.* Physica D **29** (1987), 95–127.

[Fen79] N. Fenichel, *Geometric singular perturbation theory for ordinary differential equations*, J. Diff. Eq., **31** (1979), 53–98.

[FB85] R.J. Field and M. Burger. *Oscillations and Travelling Waves in Chemical Systems.* Wiley, New York, 1985.

[FS90] E. Fontich and C. Simo. *Invariant manifolds for near identity differentiable maps and splitting of separatrices.* Ergodic Theory Dyn. Syst. **10** (1990), 319–346.

[GM98] T. Gallay and A. Mielke. *Diffusive mixing of stable states in the Ginzburg-Landau equation.* Comm. Math. Physics **199** (1998), 71–97.

[Gel97] V. Gelfreich. *Reference systems for splitting of separatrices.* Nonlinearity **10** (1997), 175–193.

[GNN97] B. Gidas, W.-M. Ni, and L. Nirenberg. *Symmetry of positive solutions of nonlinear elliptic equations in* R^n. Adv. Math., Suppl. Stud. **7A** (1981), 369–402.

[GL] L.Y. Glebsky and L.M. Lerman. *Instability of small stationary localized solutions to a class of reversible* $1+1$ *PDEs.* Nonlinearity **10**(1997), 389–407.

[GSS88] M. Golubitsky, I. Stewart, and D.G. Schaeffer. *Singularities and Groups in Bifurcation Theory. Volume II.* Applied Mathematical Sciences **69**, Springer-Verlag, New York, 1988.

[Gre78] J.M. Greenberg. *Axi-symmetric, time-periodic solutions of reaction-diffusion equations.* SIAM J. Appl. Math. **34** (1978), 391–397.

[GH] J. Guckeneimer and P. Holmes. *Nonlinear oscillations, dynamical systems, and bifurcations of vector fields.* Applied Mathematical Sciences **42**, Springer-Verlag, New York, 1990.

[CH] S.-N. Chow and J. Hale. *Methods of Bifurcation Theory.* Grundlehren der Mathematischen Wissenschaften **251**, Springer-Verlag, New York-Berlin, 1982.

[Hag81] P.S. Hagan. *Target patterns in reaction-diffusion systems.* Adv. Appl. Math. **2** (1981), 400–416.

[Hag82] P.S. Hagan. *Spiral waves in reaction-diffusion equations.* SIAM J. Appl. Math. **42** (1982), 762–786.

[HK95] M. Haragus and K. Kirchgässner. *Breaking the dimension of a steady wave: Some examples.* In Doelman, A. (ed.) et al., Nonlinear dynamics and pattern formation in the natural environment. Proceedings of the international conference held in Noordwijkerhout, The Netherlands, July 4-7, 1994. Pitman Res. Notes Math. Ser. **335** (1995), 119–129.

[HK96] M. Haragus and K. Kirchgässner. *Breaking the dimension of solitary waves.* In Chipot, M. (ed.) et al., Progress in partial differential equations: the Metz surveys 4. Proceedings of the conference given at the University of Metz, France during the 1994-95 'Metz Days'. Pitman Res. Notes Math. Ser. **345** (1996), 216–228.

[HS99] M. Haragus-Courcelle and G. Schneider. *Bifurcating fronts for the Taylor-Couette problem in infinite cylinders.* Z. Angew. Math. Phys. **50** (1999), 120–151.

[HSS99] M. van Hecke, C. Storm, and W. van Saarloos. *Sources, sinks and wavenumber selection in coupled CGL equations and implications for counter-propagating wave systems.* Physica D **134** (1999), 1–47.

[Hen81] D. Henry. *Geometric theory of semilinear parabolic equations.* Lecture Notes in Mathematics **840**, Springer-Verlag, Berlin, 1981.

[Hop43] E. Hopf. *Abzweigung einer periodischen Lösung von einer stationären Lösung eines Differentialsystems.* Ber. Verh. Sächs. Akad. Wiss. Leipzig, Math.-Naturw. Kl. **95** (1943), 3–22.

[HK77] L.N. Howard and N. Kopell. *Slowly varying waves and shock structures in reaction-diffusion equations.* Studies in Appl. Math. **56** (1977), 95–145.

[IA92] G. Iooss and M. Adelmeyer. *Topics in Bifurcation Theory and Applications.* Advanced Series in Nonlinear Dynamics **3**, World Scientific, Singapore (1992).

[IK92] G. Iooss and K. Kirchgässner. *Water waves for small surface tension: an approach via normal form.* Proc. Roy. Soc. Edinburgh A **122** (1992), 267–299.

[IM91] G. Iooss and A. Mielke. *Bifurcating time–periodic solutions of Navier–Stokes equations in infinite cylinders.* J. Nonlinear Science **1** (1991), 107–146.

[IP93] G. Iooss and Pérouème. *Perturbed homoclinic solutions in reversible 1:1 resonance vector fields.* J. Differ. Equations **102** (1993) 62–88.

[KR00] T. Kapitula and J. Rubin. *Existence and stability of standing hole solutions to the complex Ginzburg-Landau equations.* Nonlinearity **13** (2000), 77–112.

[KS98] T. Kapitula and B. Sandstede. *Stability of bright solitary wave solutions to perturbed nonlinear Schrödinger equations.* Physica D **124** (1998), 58–103.

[KH95] A. Katok and B. Hasselblatt. *Introduction to the Modern Theory of Dynamical Systems.* Encyclopedia of Mathematics and its Applications, **54**. Cambridge University Press, Cambridge, 1995.

[Kel67] A. Kelley. *The stable, center-stable, center, center-unstable, unstable manifolds.* J. Differ. Equations **3** (1967), 546–570.

[Kir82] K. Kirchgässner. *Wave solutions of reversible systems and applications.* J. Differ. Equations **45** (1982), 113–127.

[Kir88] K. Kirchgässner. *Nonlinear resonant surface waves and homoclinic bifurcation.* Adv. Appl. Mech. **26** (1988), 135–181.

[KR96] K. Kirchgässner and G. Raugel. *Stability of fronts for a KPP-system – the noncritical case.* In G. Dangelmayr et al., Dynamics of nonlinear waves in dissipative systems: reduction, bifurcation and stability. Pitman Res. Notes Math. Ser. **352** (1996), 147–208.

[KPP37] A. Kolmogorov, I. Petrovsky, and N. Piscounov. *Etude de l'équation de la diffusion avec croissance de la quantité de matière et son application à un problème biologique.* Moscow Univ. Math. Bull. **1** (1937), 1–25.

[KH81a] N. Kopell and L.N. Howard. *Target patterns and horseshoes from a perturbed central-force problem: some temporally periodic solutions to reaction-diffusion equations.* Studies in Appl. Math. **64** (1981), 1–56.

[KH81b] N. Kopell and L.N. Howard. *Target patterns and spiral solutions to reaction-diffusion equations with more than one space dimension.* Adv. Appl. Math. **2** (1981), 417–449.

[KP97] I. Kuzin and S. Pohozaev. *Entire Solutions of Semilinear Elliptic Equations.* Prog. Nonlin. Differ. Equations and their Applications **33**, Birkhäuser, Basel, 1997.

[Lin90] X.-B. Lin. *Using Melnikov's method to solve Shilnikov's problems.* Proc. R. Soc. Edinburgh **116A** (1990) 295–325.

[Lom00] E. Lombardi. *Oscillatory integrals and phenomena beyond all algebraic orders with applications to homoclinic orbits in reversible systems.* Lecture Notes in Mathematics **1741**, Springer-Verlag, Berlin, 2000.

[LPSS00] G. Lord, D. Peterhof, B. Sandstede, and A. Scheel. *Numerical computation of solitary waves of elliptic equations in infinite cylinders.* SIAM J. Numerical Analysis **37** (2000), 1420–1454.

[MSW94] J.C. van der Meer, J.A. Sanders, and A. Vanderbauwhede. *Hamiltonian structure of the reversible nonsemisimple 1:1 resonance.* In Chossat, P. (ed.), Dynamics, bifurcation and symmetry. New trends and new tools. Proceedings of the NATO Advanced Research Workshop, E. B. T. G. Conference, Cargèse, France, September 3-9, 1993. NATO ASI Ser., Ser. C, Math. Phys. Sci. **437** (1994), 221–240.

[Mie86] A. Mielke. *A reduction principle for nonautonomous systems in infinite-dimensional spaces.* J. Differ. Equations **65** (1986), 68–88.

[Mie88a] A. Mielke. *Reduction of quasilinear elliptic equations in cylindrical domains with applications.* Math. Meth. Appl. Sci. **10** (1988), 51–66.

[Mie88b] A. Mielke. *Saint Venant's problem and semi–inverse solutions in nonlinear elasticity.* Arch. Rat. Mech. Anal. **102** (1988), 205–229.

[Mie92] A. Mielke. *Reduction of PDEs on domains with several unbounded directions: A first step towards modulation equations.* Z. Angew. Math. Phys. **43** (1992), 449–470.

[NW69] A. Newell and J.A. Whitehead. *Finite bandwidth, finite amplitude convection.* J. Fluid Mech. **38** (1969), 279–303.

[Ni98] W.-M. Ni. *Diffusion, cross-diffusion, and their spike-layer steady states.* Notices AMS **45** (1998), 9–18.

[PSS97] D. Peterhof, B. Sandstede, and A. Scheel. *Exponential dichotomies for solitary-wave solutions of semilinear elliptic equations on infinite cylinders.* J. Differ. Equations **140** (1997), 266–308.

[Pli64] V.A. Pliss. *The reduction principle in the theory of stability of motion.* Sov. Math., Dokl. **5** (1964), 247–250. Translation from Dokl. Akad. Nauk SSSR **154** (1964), 1044–1046.

[Pom86] Y. Pomeau. *Front motion, metastability and subcritical bifurcations in hydrodynamics.* Physica D **23** (1986), 3–11.

[PZM85] Y. Pomeau, S. Zaleski, and P. Manneville. *Axisymmetric cellular structures revisited.* Z. Angew. Math. Phys. **36** (1985), 367–394.

[RK98] G. Raugel and K. Kirchgässner. *Stability of fronts for a KPP-system, II: The critical case.* J. Differ. Equations **146** (1998), 399–456.

[RK00] H. Riecke and L. Kramer. *The stability of standing waves.* Physica D **137** (2000), 124–142.

[vSH92] W. van Saarloos and P.C. Hohenberg. *Fronts, pulses, sources and sinks in generalized complex Ginzburg-Landau equations.* Physica D **56** (1992), 303–367.

[San93] B. Sandstede. *Verzweigungstheorie homokliner Verdopplungen.* Doctoral thesis, University of Stuttgart, 1993.

[SS99] B. Sandstede and A. Scheel. *Essential instability of pulses and bifurcations to modulated travelling waves.* Proc. Roy. Soc. Edinburgh. A **129** (1999), 1263–1290.

[SS00a] B. Sandstede and A. Scheel. *Gluing unstable fronts and backs together can produce stable pulses.* Nonlinearity **13** (2000), 1465–1482.

[SS00b] B. Sandstede and A. Scheel. *Absolute and convective instabilities of waves on unbounded and large bounded domains.* Physica D **145** (2000), 233–277.

[SS00c] B. Sandstede and A. Scheel. *Essential instabilities of fronts: bifurcation and bifurcation failure.* Dynamical Systems: An International Journal **16** (2001), 1–28.

[SS00d] B. Sandstede and A. Scheel. *On the structure of spectra of modulated travelling waves.* Mathematische Nachrichten **232** (2001), 39–93.

[SS00e] B. Sandstede and A. Scheel. *On the stability of periodic travelling waves with large spatial period.* J. Differ. Equations, **172** (2001), 134–188.

[Sche98] A. Scheel. *Bifurcation to spiral waves in reaction-diffusion systems.* SIAM J. Math. Anal. **29** (1998), 1399–1418.

[Schn95] G. Schneider. *Validity and limitation of the Newell-Whitehead equation.* Math. Nachr. **176** (1995), 249–263.

[Shu87] M. Shub. *Global Stability of Dynamical Systems.* Springer-Verlag, New York, 1987.

[Stu97] C.A. Stuart. *Bifurcation from the essential spectrum.* In Matzeu, Michele (ed.) et al., Topological nonlinear analysis II: degree, singularity and variations. Frascati, Italy, June 1995. Prog. Nonlinear Differ. Equ. Appl. **27** (1997), 397–443.

[SH77] J.B. Swift and P.C. Hohenberg. *Hydrodynamic fluctuations at the convective instability.* Phys. Rev. A **15** (1977), 319–328.

[Tak97] I. Takagi. *Spiky patterns and their stability in a reaction-diffusion system.* In Choe, H.J. (ed.) et al., Proceedings of Korea-Japan partial differential equations conference, Taejon, Republic of Korea, December 16–18, 1996. Lect. Notes Ser., Seoul **39** (1997), 1–8.

[Tho17] D.W. Thompson. *On Growth and Form.* Cambridge University Press, 1917.

[Tur52] A. Turing. *The chemical basis of morphogenesis.* Phil. Trans. Roy. Soc. B **237** (1952), 37–72.

[Van89] A. Vanderbauwhede. *Center manifolds, normal forms and elementary bifurcations.* Dynamics Reported **2** (1989), 89–169.

[VI91] A. Vanderbauwhede and G. Iooss. *Center manifold theory in infinite dimensions.* Dynamics Reported, New Series **1** (1991), 125–163.

[Wat22] G.N. Watson. *Theory of Bessel functions.* Cambridge Univ. Press, 1922.

[Yos71] K. Yosida. *Functional Analysis.* Springer-Verlag, Berlin, 1971.

Editorial Information

To be published in the *Memoirs*, a paper must be correct, new, nontrivial, and significant. Further, it must be well written and of interest to a substantial number of mathematicians. Piecemeal results, such as an inconclusive step toward an unproved major theorem or a minor variation on a known result, are in general not acceptable for publication. Papers appearing in *Memoirs* are generally longer than those appearing in *Transactions*, which shares the same editorial committee.

As of June 1, 2003, the backlog for this journal was approximately 3 volumes. This estimate is the result of dividing the number of manuscripts for this journal in the Providence office that have not yet gone to the printer on the above date by the average number of monographs per volume over the previous twelve months, reduced by the number of volumes published in four months (the time necessary for preparing a volume for the printer). (There are 6 volumes per year, each containing at least 4 numbers.)

A Consent to Publish and Copyright Agreement is required before a paper will be published in the *Memoirs*. After a paper is accepted for publication, the Providence office will send a Consent to Publish and Copyright Agreement to all authors of the paper. By submitting a paper to the *Memoirs*, authors certify that the results have not been submitted to nor are they under consideration for publication by another journal, conference proceedings, or similar publication.

Information for Authors

Memoirs are printed from camera copy fully prepared by the author. This means that the finished book will look exactly like the copy submitted.

The paper must contain a *descriptive title* and an *abstract* that summarizes the article in language suitable for workers in the general field (algebra, analysis, etc.). The *descriptive title* should be short, but informative; useless or vague phrases such as "some remarks about" or "concerning" should be avoided. The *abstract* should be at least one complete sentence, and at most 300 words. Included with the footnotes to the paper should be the 2000 *Mathematics Subject Classification* representing the primary and secondary subjects of the article. The classifications are accessible from `www.ams.org/msc/`. The list of classifications is also available in print starting with the 1999 annual index of *Mathematical Reviews*. The Mathematics Subject Classification footnote may be followed by a list of *key words and phrases* describing the subject matter of the article and taken from it. Journal abbreviations used in bibliographies are listed in the latest *Mathematical Reviews* annual index. The series abbreviations are also accessible from `www.ams.org/publications/`. To help in preparing and verifying references, the AMS offers MR Lookup, a Reference Tool for Linking, at `www.ams.org/mrlookup/`. When the manuscript is submitted, authors should supply the editor with electronic addresses if available. These will be printed after the postal address at the end of the article.

Electronically prepared manuscripts. The AMS encourages electronically prepared manuscripts, with a strong preference for $\mathcal{A}_{\mathcal{M}}\mathcal{S}$-LaTeX. To this end, the Society has prepared $\mathcal{A}_{\mathcal{M}}\mathcal{S}$-LaTeX author packages for each AMS publication. Author packages include instructions for preparing electronic manuscripts, the *AMS Author Handbook*, samples, and a style file that generates the particular design specifications of that publication series. Though $\mathcal{A}_{\mathcal{M}}\mathcal{S}$-LaTeX is the highly preferred format of TeX, author packages are also available in $\mathcal{A}_{\mathcal{M}}\mathcal{S}$-TeX.

Authors may retrieve an author package from e-MATH starting from `www.ams.org/tex/` or via FTP to `ftp.ams.org` (login as `anonymous`, enter username as password, and type `cd pub/author-info`). The *AMS Author Handbook* and the *Instruction Manual* are available in PDF format following the author packages link from `www.ams.org/tex/`. The author package can be obtained free of charge by sending email to `pub@ams.org` (Internet) or from the Publication Division, American Mathematical Society, 201 Charles St., Providence, RI 02904, USA. When requesting an author package, please specify AMS-LaTeX or AMS-TeX, Macintosh or IBM (3.5) format, and the publication in which your paper will appear. Please be sure to include your complete mailing address.

Sending electronic files. After acceptance, the source file(s) should be sent to the Providence office (this includes any TeX source file, any graphics files, and the DVI or PostScript file).

Before sending the source file, be sure you have proofread your paper carefully. The files you send must be the EXACT files used to generate the proof copy that was accepted for publication. For all publications, authors are required to send a printed copy of their paper, which exactly matches the copy approved for publication, along with any graphics that will appear in the paper.

TeX files may be submitted by email, FTP, or on diskette. The DVI file(s) and PostScript files should be submitted only by FTP or on diskette unless they are encoded properly to submit through email. (DVI files are binary and PostScript files tend to be very large.)

Electronically prepared manuscripts can be sent via email to `pub-submit@ams.org` (Internet). The subject line of the message should include the publication code to identify it as a Memoir. TeX source files, DVI files, and PostScript files can be transferred over the Internet by FTP to the Internet node `e-math.ams.org` (130.44.1.100).

Electronic graphics. Comprehensive instructions on preparing graphics are available at `www.ams.org/jourhtml/graphics.html`. A few of the major requirements are given here.

Submit files for graphics as EPS (Encapsulated PostScript) files. This includes graphics originated via a graphics application as well as scanned photographs or other computer-generated images. If this is not possible, TIFF files are acceptable as long as they can be opened in Adobe Photoshop or Illustrator. No matter what method was used to produce the graphic, it is necessary to provide a paper copy to the AMS.

Authors using graphics packages for the creation of electronic art should also avoid the use of any lines thinner than 0.5 points in width. Many graphics packages allow the user to specify a "hairline" for a very thin line. Hairlines often look acceptable when proofed on a typical laser printer. However, when produced on a high-resolution laser imagesetter, hairlines become nearly invisible and will be lost entirely in the final printing process.

Screens should be set to values between 15% and 85%. Screens which fall outside of this range are too light or too dark to print correctly. Variations of screens within a graphic should be no less than 10%.

Inquiries. Any inquiries concerning a paper that has been accepted for publication should be sent directly to the Electronic Prepress Department, American Mathematical Society, 201 Charles St., Providence, RI 02904, USA.

Editors

This journal is designed particularly for long research papers, normally at least 80 pages in length, and groups of cognate papers in pure and applied mathematics. Papers intended for publication in the *Memoirs* should be addressed to one of the following editors. In principle the Memoirs welcomes electronic submissions, and some of the editors, those whose names appear below with an asterisk (*), have indicated that they prefer them. However, editors reserve the right to request hard copies after papers have been submitted electronically. Authors are advised to make preliminary email inquiries to editors about whether they are likely to be able to handle submissions in a particular electronic form.

Algebraic geometry to DAN ABRAMOVICH, Department of Mathematics, Boston University, 111 Cummington St., Boston, MA 02215; email: `abramovic@bu.edu`

Algebraic topology and cohomology of groups to STEWART PRIDDY, Department of Mathematics, Northwestern University, 2033 Sheridan Road, Evanston, IL 60208-2730; email: `priddy@math.nwu.edu`

Combinatorics and Lie theory to SERGEY FOMIN, Department of Mathematics, University of Michigan, Ann Arbor, Michigan 48109-1109; email: `fomin@umich.edu`

Complex analysis and complex geometry to DUONG H. PHONG, Department of Mathematics, Columbia University, 2990 Broadway, New York, NY 10027-0029; email: `phong@math.columbia.edu`

*__Differential geometry and global analysis__ to LISA C. JEFFREY, Department of Mathematics, University of Toronto, 100 St. George St., Toronto, ON Canada M5S 3G3; email: `jeffrey@math.toronto.edu`

Dynamical systems and ergodic theory to ROBERT F. WILLIAMS, Department of Mathematics, University of Texas, Austin, Texas 78712-1082; email: `bob@math.utexas.edu`

*__Geometric analysis__ to TOBIAS COLDING, Courant Institute, New York University, 251 Mercer St., New York, NY 10012; email: `colding@cims.nyu.edu`

Harmonic analysis to ALEXANDER NAGEL, Department of Mathematics, University of Wisconsin, 480 Lincoln Drive, Madison, WI 53706-1313; email: `nagel@math.wisc.edu`

Harmonic analysis, representation theory, and Lie theory to ROBERT J. STANTON, Department of Mathematics, The Ohio State University, 231 West 18th Avenue, Columbus, OH 43210-1174; email: `stanton@math.ohio-state.edu`

Number theory to HAROLD G. DIAMOND, Department of Mathematics, University of Illinois, 1409 W. Green St., Urbana, IL 61801-2917; email: `diamond@math.uiuc.edu`

*__Ordinary differential equations, and applied mathematics__ to PETER W. BATES, Department of Mathematics, Michigan State University, East Lansing, MI 48824-1027; email: `peter@math.msu.edu`

*__Partial differential equations__ to PATRICIA E. BAUMAN, Department of Mathematics, Purdue University, West Lafayette, IN 47907-1395' email: `bauman@math.purdue.edu`

*__Probability and statistics__ to KRZYSZTOF BURDZY, Department of Mathematics, University of Washington, Box 354350, Seattle, Washington 98195-4350; email: `burdzy@math.washington.edu`

*__Real analysis and partial differential equations__ to DANIEL TATARU, Department of Mathematics, University of California, Berkeley, Berkeley, CA 94720; email: `tataru@math.berkeley.edu`

All other communications to the editors should be addressed to the Managing Editor, WILLIAM BECKNER, Department of Mathematics, University of Texas, Austin, TX 78712-1082; email: `beckner@math.utexas.edu`.

Titles in This Series

787 **Michael Cwikel, Per G. Nilsson, and Gideon Schechtman,** Interpolation of weighted Banach lattices/A characterization of relatively decomposable Banach lattices, 2003

786 **Arnd Scheel,** Radially symmetric patterns of reaction-diffusion systems, 2003

785 **R. R. Bruner and J. P. C. Greenlees,** The connective K-theory of finite groups, 2003

784 **Desmond Sheiham,** Invariants of boundary link cobordism, 2003

783 **Ethan Akin, Mike Hurley, and Judy A. Kennedy,** Dynamics of topologically generic homeomorphisms, 2003

782 **Masaaki Furusawa and Joseph A. Shalika,** On central critical values of the degree four L-functions for GSp(4): The Fundamental Lemma, 2003

781 **Marcin Bownik,** Anisotropic Hardy spaces and wavelets, 2003

780 **S. Marmi and D. Sauzin,** Quasianalytic monogenic solutions of a cohomological equation, 2003

779 **Hansjörg Geiges,** h-principles and flexibility in geometry, 2003

778 **David B. Massey,** Numerical control over complex analytic singularities, 2003

777 **Robert Lauter,** Pseudodifferential analysis on conformally compact spaces, 2003

776 **U. Haagerup, H. P. Rosenthal, and F. A. Sukochev,** Banach embedding properties of non-commutative L^p-spaces, 2003

775 **P. Lochak, J.-P. Marco, and D. Sauzin,** On the splitting of invariant manifolds in multidimensional near-integrable Hamiltonian systems, 2003

774 **Kai A. Behrend,** Derived ℓ-adic categories for algebraic stacks, 2003

773 **Robert M. Guralnick, Peter Müller, and Jan Saxl,** The rational function analogue of a question of Schur and exceptionality of permutation representations, 2003

772 **Katrina Barron,** The moduli space of $N = 1$ superspheres with tubes and the sewing operation, 2003

771 **Shigenori Matsumoto,** Affine flows on 3-manifolds, 2003

770 **W. N. Everitt and L. Markus,** Elliptic partial differential operators and symplectic algebra, 2003

769 **Jie Wu,** Homotopy theory of the suspensions of the projective plane, 2003

768 **R. Höpfner and E. Löcherbach,** Limit theorems for null recurrent Markov processes, 2003

767 **Po Hu,** S-modules in the category of schemes, 2003

766 **Su Gao and Alexander S. Kechris,** On the classification of Polish metric spaces up to isometry, 2003

765 **Robert Bieri and Ross Geoghegan,** Connectivity properties of group actions on non-positively curved spaces, 2003

764 **J. Spandaw,** Noether-Lefschetz problems for degeneracy loci, 2003

763 **Yasuyuki Kachi and Eiichi Sato,** Segre's reflexivity and an inductive characterization os hyperquadrics, 2002

762 **Leiba Rodman, Ilya M. Spitkovsky, and Hugo Woerdeman,** Abstract band method via factorization, positive and band extensions of multivariable almost periodic matrix functions, and spectral estimation, 2002

761 **Oliver Druet and Emmanuel Hebey,** The AB program in geometric analysis : Sharp Sobolev inequalities and related problems, 2002

760 **Markus Banagl,** Extending intersection homology type invarients to non-Witt spaces, 2002

759 **Donald M. Davis,** From representation theory to homotopy groups, 2002

758 **Alan Forrest, John Hunton, and Johannes Kellendonk,** Topological invariants for projection method patterns, 2002

757 **Douglas Bowman,** q-difference operators, orthogonal polynomials, and symmetric expansions, 2002

756 **José Ignacio Cogolludo-Agustín,** Topological invariants of the complement to arrangements of rational plane curves, 2002

755 **M. A. Mandell and J. P. May,** Equivariant orthogonal spectra and S-modules, 2002

754 **Edward L. Green, Idun Reiten, and Øyvind Solberg,** Dualities on generalized Koszul algebras, 2002

753 **Daniel Panazzolo,** Desingularization of nilpotent singularities in families of planar vector fields, 2002

752 **Linus Kramer,** Homogeneous spaces, Tits buildings, and isoparametric hypersurfaces, 2002

751 **Bruce Allison, Georgia Benkart, and Yun Gao,** Lie algebras graded by the root systems BC_r, $r \geq 2$, 2002

750 **Masaki Izumi and Hideki Kosaki,** Kac algebras arising from composition of subfactors: General theory and classification, 2002

749 **Nanhua Xi,** The based ring of two-sided cells of affine Weyl groups of type $\widetilde{A}_{n-1}$, 2002

748 **Jürgen Ritter and Alfred Weiss,** The lifted root number conjecture and Iwasawa theory, 2002

747 **Armand Borel, Robert Friedman, and John W. Morgan,** Almost commuting elements in compact Lie groups, 2002

746 **Peter Niemann,** Some generalized Kac-Moody algebras with known root multiplicities, 2002

745 **Mikhail A. Lifshits and Werner Linde,** Approximation and entropy numbers of Volterra operators with application to Brownian motion, 2002

744 **Roger Chalkley,** Basic global relative invariants for homogeneous linear differential equations, 2002

743 **Heng Sun,** Spectral decomposition of a covering of $GL(r)$: the Borel case, 2002

742 **J. E. Gilbert, Y. S. Han, J. A. Hogan, J. D. Lakey, D. Weiland, and G. Weiss,** Smooth molecular functions and singular integral operators, 2002

741 **Francisco Santos,** Triangulations of oriented matroids, 2002

740 **Rick Durrett,** Mutual invadability implies coexistence in spatial models, 2002

739 **Georgios K. Alexopoulos,** Sub-Laplacians with drift on Lie groups of polynomial volume growth, 2002

738 **Yasuro Gon,** Generalized Whittaker functions on $SU(2,2)$ with respect to the Siegel parabolic subgroup, 2002

737 **Arjen Doelman, Robert A. Gardner, and Tasso J. Kaper,** A stability index analysis of 1-D patterns of the Gray-Scott model, 2002

736 **Wojciech Chachólski and Jérôme Scherer,** Homotopy theory of diagrams, 2002

735 **Martina Brück, Xi Du, Joonsang Park, and Chuu-Lian Terng,** The submanifold geometries associated to Grassmannian systems, 2002

734 **Michel Van den Bergh,** Blowing up of non-commutative smooth surfaces, 2001

733 **Milé Krajčevski,** Tilings of the plane, hyperbolic groups and small cancellation conditions, 2001

732 **Jan O. Kleppe, Juan C. Migliore, Rosa Miró-Roig, Uwe Nagel, and Chris Peterson,** Gorenstein liaison, complete intersection liaison invariants and unobstructedness, 2001

For a complete list of titles in this series, visit the AMS Bookstore at **www.ams.org/bookstore/**.